American Winescapes

American Winescapes

The Cultural Landscapes of America's Wine Country

Gary L. Peters

Routledge
Taylor & Francis Group

NEW YORK AND LONDON

Geographies of the Imagination

First published 1997 by Westview Press

Published 2018 by Routledge
605 Third Avenue, New York, NY 10017
4 Park Square, Milton Park, Abingdon, Oxon OX14 4RN

Routledge is an imprint of the Taylor & Francis Group, an informa business

A CIP catalog record for this book is available from the Library of Congress.

ISBN 13: 978-0-8133-2856-0 (pbk)
ISBN 13: 978-0-8133-2855-3 (hbk)

Design by Jane Raese

Contents

Introduction: A Geographer's Appreciation of America's Wine Country

1 Grapevines

2 Major Cultivars in American Vineyards Today

3 American Environments for Wine Grapes

4 American Wine Making Comes of Age

5 Wine Regions and Wine Labels

6 American Viticultural Landscapes

7 Seasons, Ceremonies, and Wine-Judging Events

8 The Viticultural Area as a Working Landscape

9 Communicating About Grapes and Wines

10 America's Viticultural Future

Illustrations

Acknowledgments

Authors incur many debts in the course of writing a book, and I am no exception. My special thanks go first to my wife and children—Carol, Jason, and Erica. Over the years they have endured more than their share of visits to vineyards, wineries, and wine towns; more recently they left me to the task of writing this book with no complaints about the time it took away from them.

I also owe considerable thanks to three geography colleagues: Judith Tyner, who produced the maps for this book on very short notice, and both James Curtis and Nicholas Polizzi, who read and commented on earlier versions of the manuscript. Without their help the book would have suffered in a number of ways.

Finally, I would like to thank the many people at wineries across the nation who kindly responded to my letters, phone calls, and personal visits. Their help is greatly appreciated. In addition, the following individuals provided help that required more of their time and effort, and I offer them my heartfelt thanks: Cynthia Hill, Gideon Beinstock, and his wife, Saron (Renaissance Vineyard and Winery); Harvey Posert (Robert Mondavi Winery); Christopher Reed and Marty Laplante (Benziger Family Winery); Leon Santoro (Orfila Vineyards); Hector Bedolla (Hambrecht Vineyards); Jo Diaz (Belvedere Winery); Tom Levesque (Wente Brothers); Wendell C.M. Lee (Wine Institute); and Perky Ramroth (Bureau of Alcohol, Tobacco, and Firearms).

Of course, though I have had considerable help along the way, neither the individuals noted above, nor anyone other than I, can be held responsible for whatever shortcomings this book might have.

Gary L. Peters

$\mathscr{I}$ntroduction:
A Geographer's Appreciation
of America's Wine Country

$\mathcal{I}$MAGINE YOURSELF SITTING in the restaurant veranda at the Domaine Chandon winery in California's ethereal Napa Valley on a warm September evening. The powerful, penetrating light of a late summer day softens around you. The sun slips quietly behind the Mayacamas Mountains; shadows of oaks and conifers grow longer, gradually engulfing you. Across the valley, the Vaca Mountains turn rufescent, then become even redder in the waning light. Throughout the valley, the hectic harvest pace slackens as weary workers seek an evening of quiet and rest—lights are on late at local wineries as fresh grapes begin their journey toward becoming wine. Wine makers fuss about, anxiously smelling and tasting the grapes, making sure that the tangle of hoses is properly routed, praying for another good year. Meanwhile, your waitress brings your cool glass of sparkling Blanc de Noirs, reminds you of the evening's specials, then leaves you to decide what to eat, to linger quietly and savor the gentle coming of the night.

You're in "wine country," and in all likelihood you're loving every minute of it. You came here because you knew, or at least had some idea, what to expect—a dramatic and welcome change of *place* (and probably pace as well!). You are, like most everyone else, a geographer at heart, even though you might never have thought about it. Places, locations, regions—most of us are intuitively interested in them. Professional geographers—and I am one—simply carry those interests further, into something of a passion, perhaps. They seek to understand more about places—who lives and works in them, how their landscape features have been shaped over time, how they differ from other places, and how they continue to evolve.

"Wine country," as the term is used by typical travelers and wine lovers, sets apart in their minds places that are characterized by the presence of vineyards, wineries, and often small towns that serve the local population and visitors as well. These are working landscapes, but to many they seem to offer much more. At their richest, they can be synonymous with civilized enjoyment; food, wine, and conversation often come together here in harmonious ways. As Robert Mondavi wrote (Meyer 1989:6):

> We believe wine is the temperate, civilized, sacred, romantic mealtime beverage recommended in the Bible. It is a liquid food that has been part of civilization for 8,000 years. Wine has been praised for centuries by statesmen, scholars, poets, and philosophers. It has been used as a re-

ligious sacrament, as the primary beverage of choice for food, and as a source of pleasure and diversion.

In contrast to whatever romantic notions about the wine industry we might hold, however, we should also add that wine growing is a serious business. It is currently estimated to have an annual value of more than $12 billion in the United States, which ranks fifth in production among wine-growing nations behind Italy, France, Argentina, and Spain, in that order. In their excellent study of the economics of American wine making, journalists Jay Stuller and Glen Martin (1994:5) have put it more bluntly, telling us that "the American wine industry is a hurly-burly venue that includes cutthroat financial dealings and brutal competition for sales."

Americans have long had a fondness for agricultural landscapes. Although for some people rural landscapes are still reminiscent of the Jeffersonian ideal of an agrarian republic, most see them more realistically as an occasional refuge from the bustling life of modern American cities. To wander among landscapes permeated with colorful crops, unusual buildings, strange equipment, small country roads, and fences is to find respite from the city—from its accelerated pace, its congestion, its noises and smells, its congested center and dispersed suburbs.

Wine and Geography

The study of wine and geography constitutes a delightful marriage, a union not likely to end in either disillusion or dissolution.

Professional geographers approach their studies of wine in two separate, if not always distinctively differentiated, ways. One approach is regional and the other, topical; the two approaches come together at times, each helping us to better comprehend the other.

For most geographers, the topical study of wine would probably begin with maps that show where wine grapes are grown and where wines are made. These maps are descriptive and answer the basic "where" questions for us; they show us the spatial distribution of wine growing (a term used to include both viticulture, or the growing of grapes, and enology, or the making of wines). These maps then raise other questions, especially the following one: Why is wine growing located where it is? Answering such "why" questions leads us in search of related variables, from the annual amount of rainfall and temperatures during the growing season to the market for wines of different types, which is itself a reflection of various cultural characteristics.

The places where grape growing and wine making are found together delineate discrete regions, which we could simply call "wine regions," or "wine country." Such regions are characterized by the presence of one or more specific criteria, such as the presence of vineyards and wineries, which give the region its distinctiveness. From the topical maps, then, we could identify the locations of a nation's wine regions, just as we could identify its steel-producing regions, wheat-growing regions, or urban regions.

Wine regions most certainly are distinctive; they differ both from other types of agricultural regions and from urban or manufacturing regions. In turn, their uniqueness generates considerable appeal, hence the consequential popularity of visits to wine country. Furthermore, as geographer James Newman (1986:301) once commented: "The geography of wine does not end with a landscape. Color, smell, and taste of wine, including judgments about quality, most often stamp a region with its identity." We don't hear many people talk about visiting "beer country," for example—fields of barley and hops are usually grown far from where the brewers ply their trade, and those fields lack the grace and beauty of long rows of well-tended vines, bright green beneath the summer sun and heavily laden with ripe grapes as fall approaches. At the same time, however, though they have common attributes, not all "winescapes" are the same.

Wine regions may be viewed along a continuum, from landscapes in which wine growing is virtually the only agricultural enterprise to those in which wine grapes are only thinly intermixed with numerous other crops, including apples, cherries, peaches, prunes, walnuts, and even berries. At one end of the continuum is California's Napa County—which includes America's archetypal wine region, the Napa Valley—where in one recent year the total value of crop and livestock production was $154,055,000, of which wine-grape production accounted for $147,161,000, or 95.5 percent, of the county's total agricultural output (Napa County Agricultural Crop Report 1994:1, 6). The continuum extends downward from there to embrace counties in which wine growing accounts for only some small percentage of total agricultural output and commercial wineries may be altogether absent. Nonetheless, all wine regions are of interest from the standpoint of the geographer or the eager tourist willing to seriously consider the landscape as he or she passes through it. Viticultural landscapes have much to tell us.

Regions and Landscapes

The definition of a wine region can generally be agreed upon. Wine regions have boundaries that can be drawn around them, and that process has been going on unofficially for decades and officially (as implemented by the U.S. Bureau of Alcohol, Tobacco, and Firearms [BATF]) for more than a decade now, with the BATF designation of American Viticultural Areas (AVAs). The terms "place" and "landscape" are more ambiguous. The professional literature in geography is enlivened with detailed discussions of place and landscape, both of which are central to the very core of the discipline. A place can best be thought of first as a locality, a location, or a particular milieu; places, then, are considerably influenced by the people who occupy them and by the culture within which they exist. They may have a notable distinctiveness—what both geographers and many great novelists recognize as a sense of place, or local ambience. A landscape consists of what we see in front of us, from the shape of the land and its covering of vegetation to the cultural impress of roads, buildings, land uses, and inhabitants. Geographers often distinguish between natural and cultural landscapes, though there are few of the former remaining today; most landscapes sustain marks of human exploration and habitation.

Agricultural landscapes constitute specific types of cultural landscapes. They consist primarily of topographic surfaces, the kinds of plants and animals that are produced and nurtured within them, and the assemblage of structures that people have added to make these landscapes more productive. From rolling fields of grain and gleaming grain elevators to trellised rows of grapevines and adjacent wineries, agricultural landscapes reflect what people have been able to do with their natural environment. Of course, all cultural landscapes change over time, influenced by everything from the discovery of new crops and improved technology to changing tastes among consumers.

Geographers have long studied agricultural landscapes, especially their form and evolution. Descriptive studies of agricultural regions have focused on their appearance, principally on cropping practices and the array of built structures found within the regions. These structures, in turn, serve distinctive purposes. Some farmhouses, for example, may have been built at least partly to be pleasing to their inhabitants, whereas others may have been designed to facilitate certain functions. Over time, of course, as land uses change, buildings once designed for one function may be converted to another; others remain only as relics, often slumping under the weight of gravity, solemn evidence of changing times. A number of old dairy barns in the United States, for example, are now serving as wineries; horse barns in the Midwest, by contrast, have normally been allowed to deteriorate as tractors have replaced horses in the fields. Built environments not only tell us about a region's function but also provide clues to local cultural influences and even to the use of local building materials.

Agricultural landscapes evolve as the result of numerous individuals making decisions about how to use the land, how to build buildings, and how best to design settlements to serve rural residents. Confronted with similar physical environments, rational people end up making similar decisions about which crops to grow and how to grow them. Topography, climate, inherent fertility of the soil, and water availability combine to determine which plants and animals can be maintained profitably in a particular location. Some regions may be extremely constrained by environmental circumstances, whereas others have a wide array of farming possibilities. For example, a farmer in California's Great Central Valley may have a choice of a dozen or more different crops that can be grown profitably on a given piece of land, whereas a Kansas farmer may be restricted to only one crop, wheat.

Crops can be viewed as having both environmental and economic optima, and the two options might often conflict, as when demand or competition from more valuable crops may displace a given crop from its environmentally optimum location. For example, the climate of California's Great Central Valley is not only excellent for wheat and corn but also for vineyards and orchards. Most often, the result is to locate vineyards and orchards in the best locations—those with good soils and adequate water supplies—and push wheat and corn onto more marginal lands or toward locations that are more remote from major markets. This happens because peaches and grapes are more valuable and perishable than wheat or corn. Still, the crop can be elbowed only so far from its optimum before it can no longer be grown economically at all.

Where the combination of environmental conditions in a given location represents an optimum for more than one crop, the crop that produces the highest return per acre is most likely to be grown by economically rational farmers. For example, in some of California's coolest viticultural areas, grape growers typically have several varieties to choose from; the overwhelming choice in recent years, however, has been Chardonnay because of high prices and strong consumer demand. Riesling and Gewürztraminer grape growing has been declining in those same regions, mainly because these varieties have fallen out of favor among American wine consumers, not because they can't be grown successfully in the same locations.

As specialized agricultural regions evolve, the dominant activities within them produce a distinctive assemblage of common landscape features. Where livestock is present, for example, fences are common. Vineyard and orchard landscapes, however, often have few or no fences. Greenhouses and nurseries may or may not be present. The shape of fields, the flow of water, and the ways that farms are connected to the rest of the world (by roads and telephone lines, for example) are all determined to some extent by the type of agriculture that is being carried on in a region. Other important factors include whether farmers live on their land or in nearby villages and the local topography. In turn, agricultural regions usually develop systems for the distribution of their products as well; collection points, inspection stations, processing centers, packing sheds, and warehouses may be common features in the agricultural landscape.

Finally, rural landscapes in the United States are increasingly filling up with nonfarm populations, especially in agricultural regions that lie in close proximity to major metropolitan areas. Land uses in such rural landscapes are often threatened by rising land prices, increasing highway traffic, and growing conflicts between newcomers and long-time residents; the boundary between rural and urban has become increasingly blurred in modern America. California's Santa Clara Valley, for example, was once prime vineyard land, renowned for its Cabernet Sauvignon. Today, it is a landscape strewn with seemingly endless suburbs, fenced-off freeways, wide streets, parking lots, shopping centers, and a vast array of high-tech industries; it has become "Silicon Valley." Landscapes are seldom static.

Winescapes

America's viticultural landscapes, or winescapes, are human creations that have unfolded as the elements of the natural landscape—landforms, climates, vegetation, soils, and water supplies—have been brought together with the environmental needs of wine grapes. Such landscapes are worthy of attention as specific examples of agricultural landscapes, and understanding their evolution and distinctive features can add to our knowledge of rural landscapes in the United States. We can understand more about viticultural landscapes by considering the three fundamental elements that shape them: (1) the grapes and their needs, (2) the natural environments that best meet those needs, and (3) the viticulturists and wine makers who determine everything from the varieties of grapes, spacing of the vines, and trellising systems to the final product that enters the bottle. Furthermore, all of these elements can come together to produce wine regions only within the broader context of cultural practices and economic viability: If Americans drank no wine, we would probably produce little if any of it because sales would depend entirely on export markets.

This book explores why wine country exists in some places and not in others and why one can expect to find both similarities and differences among wine regions within the United States. Although the purpose of wine regions is always to produce wine, these regions may differ considerably in scale of operation, the varieties of grapes grown, the way viticulturists plant and trellis the vines, and even in irrigation practices and pest-control methods. In each case, however, a distinctive winescape is created; most such regions are capable of both generating a profit for wine growers and bringing considerable enjoyment to those who venture into them.

This book examines wines, wine growing, how viticultural landscapes have evolved in the United States, and why many of them are now so frequently visited. It is about American wines and winescapes. It begins with a historical view of wine grapes—their gradual diffusion to the United States, the native grapes that were already here, the hybrids that have been created, and the environmental requirements of the vines. Modern wine making and viticulture are then discussed, along with the identification of wine regions in the United States and the official designation of appellations of origin, which takes readers briefly into the political geography of viticulture and introduces the Bureau of Alcohol, Tobacco, and Firearms. Wine-growing landscapes are then discussed, along with perceptions of them, seasons within them, and ceremonies and festivals that help draw people to them. The last part of the book considers visits to wine regions, discusses communications related to wines and vineyards, and takes a parting look at future prospects and problems for America's wine industry and its viticultural landscapes.

1

Grapevines

Bainbridge Island
Winery

1993 WASHINGTON
dry

1995
Norton
Orange County
Virginia

HORTON
VINEYARDS

*W*INE, THE FERMENTED JUICE of ripe grapes, has been produced and consumed by humans for longer than anyone can remember. As eminent British wine historian Hugh Johnson (1994:12) once wrote, "Mankind, as we recognize ourselves, working, quarrelling, loving and worrying, comes on the scene with the support of a jug of wine." Some of our distant ancestors, whose lives preceded the earliest written accounts of people and places, found that ripe grapes, left alone in a container of some kind, began to ferment. They knew not why (the combination of natural sugars and wild yeasts encouraged fermentation to occur), may even have fretted at the apparent loss of good mature fruit, but they must certainly have liked the fermenting juices—grapey, winy, and capable of both improving the taste of foods and inducing satisfying feelings. Without much doubt, however, it was the effect of alcohol and not the allure of its savor that first made wine popular.

Perhaps as mysterious to those early imbibers as eclipses of the sun or violent volcanic eruptions, wine seems to have found its way very early into religious experiences. Then, as now, too much of this tempting beverage may have induced visions and delusions, loosening for a while the tight grip of reality. As an old Italian proverb reminds us, "One barrel of wine can work more miracles than a church full of saints."

Wine making begins in the vineyard, first with the selection of grapevines to be planted, then continues with the spacing of those vines, their training on trellises or growth as individual standing plants, their annual pruning, and finally culminates in their general care and maintenance: Good wines require healthy vines. Ampelography is the scientific study of grapevines; ampelographers describe and classify them and study their patterns of behavior.

Practically all of the world's wine grapes belong to the genus *Vitis*, which in turn is divided into about fifty species, all of which are native either to Eurasia or North America. Only one species, however, *Vitis vinifera* (often referred to as the European wine grape—*vinifera* means wine-bearing), has proved to be of paramount importance in the world's wine industry. Although some American wines are made from the grapes of other species (mainly *Vitis labrusca*) and some from hybrids (true crosses between species), the vast majority are made from *Vitis vinifera* grapes (more than 99.99 percent of the world's wines are made from grapes of this one species, according to British wine authority Jancis Robinson [1986]).

Although there are several thousand varieties of grapes in the *Vitis vinifera* family, less than one hundred of them are important in American viticulture today in the 1990s, and most wines produced in the United States are made from far fewer cultivars (short for cultivated varieties) still. But before looking more closely at the major cultivars that dominate today's American winescapes, a few words need to be said about how *Vitis vinifera* made its way from a somewhat isolated beginning to being the dominant species in the world's vineyards as we approach the millennium's end.

Vitis Vinifera: Its Origin and Diffusion

Geographers have long been interested in where things originated and how they have dispersed. The earliest predecessors of *Vitis vinifera* evolved perhaps 60 million years ago, soon after the extinction of the dinosaurs. Spreading and differentiating, grape species became ever more numerous; their appreciation by humans, however, appears only quite recently on the geological clock (so far as we know), following the end of a series of cold surges (more than thirty glacial episodes), known collectively as the Late Cenozoic Ice Age, that gripped the earth during the last 2.5 to 3 million years or so (including all of what geologists refer to as the Pleistocene epoch and a chunk of the late Pliocene epoch as well).

The latest of this series of glacial episodes, called the Late Wisconsin in North America and the Würm in Europe, gradually ended about 11,000 years ago. We can only imagine the great continental glaciers that covered most of northern Europe and northern America, as well as the extensive glaciers that were slowly slipping down valleys from the Alps and other European and American mountain systems, slowly retreating, causing sea levels to rise and local climates to change as planetary temperatures sluggishly warmed. Glacial advances destroyed vegetation (including early species of grapes) across most of Europe and much of North America, though vines and other plants found room to survive in the southern reaches of what would later become the United States.

As the ice retreated, however, vegetation began to develop again at its margins, putting roots down in naked soils that had only recently emerged from below fields of ice, some of which had been hundreds, sometimes even thousands, of feet thick. Plants gradually worked their way both generally northward and locally up the slopes of previously glaciated valleys in mountainous areas.

Continued warming encouraged more plants, including some early members of *Vitis vinifera*, to flourish, especially in sunny sites. Although we may never know this with certainty, evidence strongly suggests that one subspecies, *Vitis vinifera sylvestris*, established a foothold on south-facing slopes of the Caucasus Mountains sometime after the last glacial episode came haltingly to a close. The Caucasus Mountains stretch eastward from the Black Sea to the Caspian Sea; the southern Russian border runs through them today, separating Russia from neighboring Georgia and Azerbaijan. Somewhere in these latter two nations, or perhaps in adjacent Armenia, early wine making began, probably between 7,000 and 9,000 years ago. Geographer Harm de Blij has suggested that the earliest grapevines to be cultivated were those of *Vitis vinifera sativa*, a plant that wine historian Hugh Johnson now argues was in fact not a subspecies at all but the first real cultivar. Grape pips (seeds) from these early cultivated vines have been found in the Transcaucasian region and have been dated to somewhere between 7000 and 5000 B.C. In *Vintage: The Story of Wine*, Hugh Johnson (1989:18) wrote the following about this early time in the history of the vine:

> To put this era of human history in some sort of perspective, it was when advanced cultures, in Europe and the Near East, had changed from a nomadic to a settled way of life and started farming as well as hunting, when speech and language reached the point where "sustained conversation was possible and the invention of writing only a matter of time," when technology was moving from stone implements to copper ones, and just about the time when the first pottery was made, in the neighbourhood of the Caspian Sea. It seems, from what faint traces we can see, that it was a peaceful time, which has left us images of fertility rather than power and conquest.

Although we will never be certain, it seems possible that wine—that newly discovered and enthusiastically received beverage—played some role in the gradual transition in human society and culture that was occurring in the lands south of the Caucasus at that time. Thucydides, for example, suggested that "the peoples of the Mediterranean began to emerge from barbarism when they learnt to cultivate the olive and the vine" (cited in Johnson 1989:35).

Recently, archaeologists have found evidence of intentional and systematic wine making that dates back to 5400 B.C.—only a millennium or two after humans first began to live in fixed houses and to till the soil. This discovery of wine residue in a 2.5-gallon jug found in a home in Hajji Firuz Tepe in western Iran (then occupied by early Sumerians) by archaeologist Mary Voigt and dated by archaeological chemist Patrick McGovern is the earliest confirmation of wine making known. The residue in the jug, which had a cap to seal it (further evidence that the vessel may originally have contained wine), was found to be primarily calcium salt of tartaric acid, which naturally occurs in high concentrations only in grapes, along with traces of a yellowish resin from the terebinth tree, which was widely used in antiquity as a means of preserving wine and slowing the growth of bacteria that could turn the wine to vinegar. Evidence of beer residue was found in the same house, suggesting that at least 7,400 years ago, some of our early ancestors were imbibing alcohol. It is not unlikely that these early folks, like people today, were seeking ways to alleviate the stresses associated with living, trying to cope with uncertainty and change.

Vitis Vinifera *in Western Europe and North Africa*

From a broad center of early domestication in the Transcaucasian region, cultivars of *Vitis vinifera* began journeys in at least three different directions—eastward and southeastward into Asia, southwestward into Egypt, and westward through Anatolia to Greece. The diffusion of viticulture through Asia carried vines into the Zagros Mountains, then on to India, China, and finally Japan; the remaining two diffusion paths gradually set the stage for *Vitis vinifera*'s appearance in North America, where numerous other species of *Vitis* were already growing, though to our knowledge none had been used by early inhabitants for wine making.

Grapevines in Egypt

By about 6,000 years ago, cultivars of *Vitis vinifera* were probably being grown by Mesopotamians (who certainly were consuming wines), and the vines were gradually dispersing southwestward toward Egypt, though nothing significant is known about the details of this journey. By about 5,000 years ago, the Babylonians were exporting wine to Egypt, and within the following millennium, grapevines were being cultivated in Egypt as well, where the consumption of wine and its association with royalty and religious ceremonies had become well established. Around 3200 B.C., Egypt became unified under the Menes people, who created the First Dynasty; kings from that dynasty were known to have had large wine cellars. Without doubt, those intrepid traders, the Phoenicians, also played a role in moving both vines and wines around the eastern end of the Mediterranean in the following millennium.

Although we know little about wine growing in Mesopotamia, Egyptian artists left paintings and decorative pottery that clearly depicted wine-making processes. The early Dynastic period in Egypt was followed by the Old Kingdom (2700–2180 B.C.), a period characterized by rulers such as Imhotep and Cheops. Both art and architecture flourished early in the Old Kingdom, culminating in the building of the great pyramids at Giza, the largest of which is Cheops (2300 B.C.). Wine was commonly consumed and produced in Egypt by this time, primarily by the wealthy, royalty, and the priesthood.

The Egyptians may have been the first people to take wine growing seriously; they studied viticulture and enology, created several new innovations (improved wine presses, for example), yet gradually turned more toward beer as their beverage of choice for everyday consumption. Although we don't know what cultivars grew in vineyards along the Nile delta, we do know from Egyptian artists that people trained the vines, stomped the grapes by foot, used clay jars for fermentation and storage, and made wines that were certainly appreciated by the nobility. We know little about the quality of these wines, but it is hard not to agree with Hugh Johnson's (1989:30) observation that we should not "dismiss what people of such culture as the Egyptian aristocracy described as good, very good or excellent, took such trouble in making and pleasure in drinking."

Vines Arrive in Europe

Although the Egyptians gradually lost interest in wine—first because of the shift to beer as a beverage of choice and then with the arrival of Islam—and today make very little of it, they seem to have played an important role in the early diffusion of *Vitis vinifera* to Europe. Between about 3000 and 2000 B.C., viticulture arrived in Crete, quite possibly from both Egypt and Anatolia (now Turkey). Wine consumption quickly became associated on this sunny Mediterranean isle with festivals, burials, and various religious events.

During the next millennium, grapevines and viticultural practices arrived in Greece as well, most likely from both Crete and Anatolia. Trade between Crete's Minoans and Greece's Mycenaeans was well established by then, and cultural practices were being transferred from Crete to the Greek peninsula, where the population was growing. Worship of Dionysus or Bacchus, both names for the Greek god of wine, provided evidence of how Greeks felt about wine and its role in their lives.

Henceforth, *Vitis vinifera* was to reach its apogee in Europe. The Etruscans (in what is now Tuscany) developed viticulture, probably before the arrival of the Greeks. Within the last thousand years before the birth of Christ, the Greeks carried vines to Sicily, Italy (which they called Enotria—the land of the vine), southern France, and the Iberian peninsula. Although the Greeks initially developed the science of ampelography, it was the Romans who were to carry viticulture to new heights and to further diffuse wine growing throughout Europe, as far northwestward as England.

By all accounts, the Romans loved their wines; bacchanalia, the Roman festival of Bacchus, celebrated the fruit of the vine with singing and dancing. Viticulture became firmly established as a significant part of the Roman agricultural base. Not only did the Romans often drink to excess, but they also were among the first to experience a number of problems that still plague the wine industry today, including cycles of overproduction, calls for protection from foreign competitors, and variations in quality (and possibly adulteration of wines as well). Roman officials may have been the first (but by no means the last) to interfere in the wine business; threatened oversupplies and falling prices were sometimes met by decrees to remove inferior vineyards, for example. Much later, governments became involved in regulating everything from production levels to the types of vines that could be grown in particular areas. The political geography of wine has a long history, some of which will be touched on later.

At the same time, the Romans developed the sciences of viticulture and enology. The growing, training, and pruning of vines were studied, along with pests that affect the vines, and viticulturists began to search more carefully for cultivars that would thrive in specific microclimates. Wooden barrels for the aging of wines were employed by Roman wine makers to improve their products, some of which were quite long-lived (though we have no idea how good they might have been). More important than those Roman innovations, however, was the continued diffusion of viticulture throughout western Europe.

Wherever the Roman legions went, viticulture seemed soon to follow. Leaving home may have been acceptable to soldiers, but leaving behind their wines must have been intolerable. *Vitis vinifera* vines were carried to France's Rhône Valley, to Bordeaux, Burgundy, the Loire Valley, and Champagne; to Germany's Rhine and Moselle Valleys; and even to merry old England (where recent decades have witnessed a considerable renewal of that nation's wine industry, mainly by using hybrid grapevines that have been developed in Germany). A look at a map of western Europe will quickly show the relationship between these Roman viticultural sites and the rivers that allowed Roman penetration into the interior of the continent. Those same rivers, of course, served to move the wine to markets, including Rome.

Within a few hundred years of the birth of Christ, Europe's viticultural map had been thoroughly drawn—and with few exceptions, its great wine regions, as we recognize them today, had already been planted with vines. Numerous cultivars had been recognized by the Romans, and the task that remained in Europe was to match specific cultivars more closely with the regions in which they would do best; in contemporary America (and even in Europe, despite strict viticultural laws) this search continues, balancing viticultural and enological knowledge with the ever-changing tastes of the world's wine consumers. In our present global economy, wine moves more widely than ever before; producers in search of growing profits chase consumers in search of value, mediated always by fluctuations in currency values and government decrees.

Vitis Vinifera *in the New World*

The history of European viticulture is interesting, but far too complex to more than sketch in for this introduction. Suffice it to say that European viticulture survived a variety of difficult periods (the Dark Ages, for example), surmounted all obstacles (including many attempts to do away with it, tariffs to discourage trade, and treaties that affected it), and somehow continued to improve technologically and qualitatively. The use of wooden barrels was raised to higher levels, especially in France; bottles and corks were introduced, Champagne was "discovered" and brought to perfection, port and sherry entered the wine lexicon, along with such enological curiosities as retsina—a Greek wine flavored with resin from the Alep pine. Wine quality in general continued to improve. (Johnson [1989] and de Blij [1983], provide more detailed accounts of this era.)

For us, however, it was the European diffusion of viticulture to the New World that deserves attention now. Soon after Columbus first arrived in America, European vines began to move across the Atlantic. The first planting of these vines occurred in the hills of Parras de la Fuente, not far from Monterrey, Mexico; Cortés encouraged settlers to plant vineyards, and the Mexican wine industry thrived. The major cultivar was the Spanish Criolla (known in Chile as Pais and in the United States as Mission, still grown in a few California vineyards today).

So successful was the Mexican wine industry that wine growers were soon exporting wines to Spain, creating one of the world's first serious "wine wars." From this successful Mexican viticultural heartland, *Vitis vinifera* vines were carried southward, to Peru, Argentina, and Chile during the sixteenth century, before an edict prohibiting new viticultural ventures was issued; viticulture moved northward as well during the same century. For example, vineyards were being established in the Rio Grande Valley before the middle of the sixteenth century (probably the earliest planted vineyards in the United States) and wines were being made from these plantings by the 1580s. At about the same time, however, Spanish winegrowers, threatened by a deluge of imported Mexican wines, sought help (in the form of import protection) from the Spanish government; heeding their pleas, in 1595 King Philip II proclaimed that there would be no more new vineyards planted in Spanish America, though his edict did not apply to Catholic religious orders.

Thus, after 1595 the Catholic orders became the only winegrowers in the Americas. By the beginning of the eighteenth century, vineyards were planted in Baja California at missions Loredo and Santo Tomás de Aquino. Although these missions were ultimately unsuccessful, later immigrants, including some Russians, sustained viticulture in Baja California's Guadalupe and Calafia Valleys, where a rejuvenated Mexican wine industry thrives today.

From Baja, vines were carried northward into Alta California during the eighteenth century. Father Junípero Serra established the first mission on California soil (San Diego de Alcala) in 1769; vineyards were planted there within the next few years. Farther north in California, another twenty missions were established, most of which also produced wines; the last in this chain of missions was completed in Sonoma in 1823. Within that same decade, settlers had also planted a few vineyards in California; for example, in 1824 Joseph Chapman planted California's first commercial vineyards, near what is now downtown Los Angeles. In 1831, Jean Louis Vignes imported a variety of new cultivars (probably the first in the state aside from the Mission grape) from Europe to plant another vineyard not far from Chapman's, and in that same decade, George Yount planted grapevines in the Napa Valley; with help from the gold rush that began in 1849, commercial viticulture in the Golden State became well established before 1900.

Early viticulture in the western United States receives most of the attention, but by the 1560s, French Huguenot pioneers in Florida were attempting to make wine (which turned out to be quite unappealing) with native species of grapes (mainly *Vitis rotundifolia*, one of many species that had already adapted to North America). In the following century, *Vitis vinifera* vines were brought from Europe and planted in Virginia, Maryland, and Pennsylvania. Native vine pests (especially the nasty phylloxera vastatrix, about which more will be said later) and severe winter weather, however, rendered these early eastern American *vinifera* experiments unsuccessful.

Native American Grapevines

By the end of the eighteenth century, viticulture in the eastern United States, especially in Pennsylvania, had become well established; hybrids, mostly developed from native grapevines rather than from *Vitis vinifera*, formed the basis of this new industry. The first important grape was known as Alexander, and over the next one hundred years, more new grapes—Catawba, Isabella, and Concord among them—added to viticultural variety in the eastern United States and allowed the industry to move into the Midwest, especially into Ohio, which only 150 years ago was being called the "American Rhine" and was then the nation's leading viticultural state. These cultivars were either members of *Vitis labrusca* (which is native to the eastern United States from the Carolinas to Maine) or crosses that included *labrusca*. Alexander, for example, is thought to have been a *labrusca-vinifera* hybrid; Concord, however, was probably a pure *labrusca*.

Concord remains the most widely planted of these varieties and is probably best known for its use in juices and jellies; next in importance today are Catawba and Niagara. As we shall see, though *Vitis vinifera* cultivars are now grown in many eastern vineyards, some of these older American hybrids, along with some newer hybrids developed by the French (primarily crosses between *vinifera* and *labrusca* vines), are still important in that region's viticulture.

As the missionaries moved northward in California, they found native American vines growing there. *Vitis girdiana* and *Vitis californica* were found tangled in riparian woodlands; attempts to make wine from them, however, were unsuccessful. Aside from those mentioned already, numerous other native American grape species grew from the Rocky Mountains eastward, though there is no evidence that Native Americans ever used them for wine production. The most widely distributed of these was *Vitis riparia*, which is found from the Rocky Mountains to the Atlantic Ocean, though not along the Gulf Coast. Other important American species include *Vitis rupestris*, *Vitis cordifolia*, and *Vitis cinerea*, though today they are important for their roots, not for their grapes.

Let's return briefly now to that ugliest of Americans, phylloxera vastatrix, a tiny aphid, sometimes called a root louse, that likes few things better than juicy *Vitis vinifera* roots. Phylloxera, which has not naturally occurred in either of the native California species, is native to American grapevines east of the Rocky Mountains, which are resistant to it as a result of long exposure. Phylloxera remained relatively unknown for a considerable time, even though it certainly was responsible for destroying most of the early *Vitis vinifera* plantings in the eastern United States. Its renown, however, stems from America's first vine exports to Europe during the middle of the nineteenth century.

Although American vines were first imported by botanists in England, quite likely to the Kew Gardens in London among other locations, it was their importation by the French that was critical for the world of wines and vines. By the 1840s, French viticulturists were having a variety of different problems with their vines, including the introduction of the fungal powdery mildew *Oidium*. Within the next decade, this fungus, thought to have originated in America, swept through the vines of France and made its way southward and east-ward, into the vineyards of Spain, Italy, eastern Europe, and even North Africa; it also brought viticulture to its knees in Ohio. How-ever, it was fortuitously discovered that Catawba and a few other American cultivars possessed at least a modicum of resistance to oid-ium. This turned out to be one of modern viticulture's worst "good news, bad news" stories.

Possible resistance to powdery mildew was the good news, of course. French viticulturists began to import American vines to re-place those sickened by powdery mildew during the 1850s, before they discovered that dusting sulfur on vines would control that de-structive fungus quite effectively. The bad news, of course, was that the tiny unlisted passengers on the imported cuttings were phyllox-era. Catastrophe followed: From around 1860, utter devastation de-scended on Europe's vineyards, and the tiny culprit was not identified until 1868. Wine writer Jancis Robinson (1986:10) has described this pest's European debut as follows :

> Phylloxera had a wonderful European *grande bouffe*, munching its way through *vinifera* roots and injecting a poisonous saliva into them. Whole vineyard areas were gradually destroyed and, thanks to the inter-national trade in *vinifera* vine cuttings, eventually blighted wine produc-tion all over Europe, in South Africa, Australia, New Zealand and even California—by then planted with *vinifera*—by the turn of the century.

Phylloxera sucks away at a vine's roots, gradually weakening the vine and leaving it susceptible to attack from other pests as well. After 1868, European viticulturists began frantically trying to find ways to control phylloxera. Vineyards were flooded in attempts to drown the devastating bugs, but with little success; chemicals were tried, but they failed, too. Precious time was being lost as phylloxera spread throughout the viticultural world, though by as early as 1870 a "cure" had actually been discovered.

Just as American vines created the problem by providing the tiny insects with a route across the Atlantic, they were also subsequently recognized as the solution. Because phylloxera was native to vines in the eastern United States, those vines had roots that had developed considerable resistance to the tiny root-sucking aphid. The key, then, was to graft *Vitis vinifera* vines onto rootstocks from native American species of vines. Studies suggested that *Vitis rotundifolia*, *Vitis rupestris*,

Vitis cinerea, and *Vitis riparia* were especially resistant to phylloxera; the transition to American rootstocks, however, was to be neither rapid nor cheap. Gradually, however, in France and throughout most of Europe, healthy grapevines again appeared in the vineyards, growing now on American roots. Today, the use of phylloxera-resistant rootstock remains the only way to control phylloxera. Since about 1980, a new form of phylloxera, known as Type B, has emerged in California's Napa Valley, where most vines had been planted on AxR#1, a rootstock that was thought to be phylloxera resistant but proved unresistant to this new strain.

The American viticultural stage has now been set; the next chapter introduces the cast of characters that play important roles in the vineyards of the United States in the mid-1990s. A look at their geographic distribution allows us to see the broad outline of the nation's viticultural map as it appears today.

2

$\mathcal{M}$ajor Cultivars in American Vineyards Today

THE MOST BASIC CHARACTERS found on America's viticultural stage, of course, are the cultivars from which we get our wine grapes. Most are members of one species, *Vitis vinifera*, though a few native American and French-American hybrids remain important, primarily in the vineyards of eastern states such as Ohio, Pennsylvania, and New York. Chardonnay, which has become the premier wine grape cultivar in the United States (Figure 2.1), is discussed first, along with other *vinifera* cultivars.

Important *Vitis Vinifera* Cultivars

In addition to those *Vitis vinifera* cultivars that are most widely grown for wine making today in the United States, several others are discussed here because they are either increasing in importance or are integral to recent wine-making trends. Viticultural examples of the latter include Italian cultivars such as Sangiovese and Nebbiolo and Rhône cultivars such as Syrah and Mourvèdre; though acreages of these grapes remain small, consumer interest has certainly been growing.

Cultivar Nobility

Among the thousands of *Vitis vinifera* cultivars, four have risen to noble status; these are recognized by wine authorities and connoisseurs as capable of producing wines that are not only of the finest quality but also of distinctive character. Two of these noble grapes, Chardonnay and Riesling, are white; the other two, Cabernet Sauvignon and Pinot Noir, are red.

Chardonnay

Often called the queen of cultivars by wine lovers, the Chardonnay grape can indeed make royal, even ethereal, wines—it is the soul of the great white Burgundies, epitomized by names such as Montrachet. According to British wine historian Hugh Johnson (1994:25), Chardonnay provides us with "firm, full, strong wine with scent and character, on chalky soils becoming almost luscious without being sweet." British wine authority Jancis Robinson (1986:106) has written

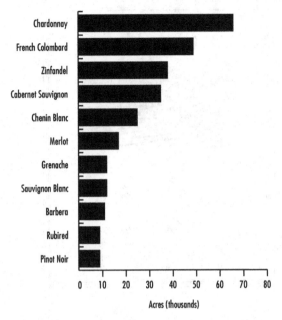

FIGURE 2.1   *California's Leading Cultivars, Mid-1990s*
Only recently has Chardonnay risen to number one in the Golden State; it is now the leading wine grape cultivar in the United States as well. *Source:* Data are from *California Grape Acreage: 1995* (Sacramento, CA: California Agricultural Statistics Service, 1996).

that "in Chardonnay is one of the happiest of all combinations: the grower loves to grow it; the winemaker loves to fashion it; and we all love to drink it."

Well-made Chardonnays are medium- to full-bodied wines that range in color from pale straw to light gold. The aroma of apples is often identified with Chardonnay, as are the aromas of peaches and even figs, on occasion. Balanced against the fruit flavors of the grape are the nuances of oak (vanilla and toastiness) that give many of the best Chardonnays their additional complexity and lushness. Unlike most white wines, Chardonnays often improve with some bottle age, though American examples are not thought of as being as long-lived as the best French white Burgundies.

Incorrectly labeled Pinot Chardonnay at one time in America (mainly in California), Chardonnay came here originally from France, where it produces the finest white wines of Burgundy and many of the fine sparkling wines of Champagne and where it is the only approved

white cultivar that can be grown. Chardonnay is the essence of such enchanted wines as Chablis, Pouilly-Fuissé, Meursault, and Bâtard-Montrachet. Without doubt, Chardonnay has found a comfortable second home on American soil—it has become *the* white wine to drink (and be seen drinking, of course!). California remains the premier producer of Chardonnay in the United States; within the Golden State, three counties in the mid-1990s each had in excess of 10,000 acres of that cultivar: Monterey (11,693), Napa (10,159), and Sonoma (12,469) (California Department of Agriculture). Chardonnay vines grow best (and produce the most prestigious wines) in the coolest viticultural regions; from the Anderson Valley in Mendocino County to the Santa Ynez Valley in Santa Barbara County, we find these great grapes in ever-increasing numbers in California.

Across America, Chardonnay has found conditions to its liking in a number of other places as well. More than 2,000 acres of Chardonnay are grown in Oregon's Willamette Valley, for example, and another 2,000 acres in neighboring Washington, primarily in the Yakima Valley. Chardonnay is also grown in smaller amounts in many other states, including Idaho, Arizona, New Mexico, Texas, Arkansas, Missouri, Ohio, Tennessee, Virginia, the Carolinas, New Jersey, Pennsylvania, Connecticut, and New York, which now has about 1,000 acres planted (mainly around the Finger Lakes and on Long Island).

Because it begins to grow early each spring, the Chardonnay vine is often subject to frost damage; it also has problems with powdery mildew and Pierce's disease, an incurable bacterial disease caused by *Xylella fastidiosa*. The disease has recently emerged as a threat to vineyards in northern California, including parts of the Napa Valley and Sonoma County. However, Chardonnay grapes (and wines made from them) command high prices, so many growers are willing and eager to grow them. Unfortunately, the temptation to grow Chardonnay at increasing distances from its optimum environment has been too great for some winegrowers. As a result, Chardonnay acreage is expanding in the delta region of California's Great Central Valley, especially around Lodi, for example, where it is likely to fall far short of its splendid potential.

More types and brands of Chardonnay wines are made in the United States today than any other varietal wines, though Cabernet Sauvignon may be a close rival. Since the 1980s, these two wines have dominated the premium table wine category in the United States, raising concerns among some American wine lovers that the variety of wines available may actually be decreasing.

Riesling (White or Johannisberg Riesling)

Let's begin by clearing up some confusion. Riesling is the eminent classical grape of Germany, where it is geographically concentrated primarily along the steeply terraced banks of the Moselle River and in the Rheingau. It is the noble grape of Schloss Vollrads and Schloss Jo-

DRY

Dr. Konstantin Frank
1994
Johannisberg Riesling

NEW YORK FINGER LAKES.
PRODUCED AND BOTTLED BY
KONSTANTIN D. FRANK & SONS VINIFERA WINE CELLARS, LTD.
HAMMONDSPORT, N.Y. 14840. ALCOHOL 11% BY VOLUME.
CONTAINS SULFITES

hannisberg (which lends its name to the variety in the United States) and a host of other famous German vineyards.

When it was transplanted elsewhere, Riesling sometimes took on other names, especially Johannisberg Riesling. The nomenclature of other grapes, such as the Sylvaner, also began to reflect use of the word Riesling in order to capitalize on the true Riesling's reputation. The Sylvaner, for example, also became known as the Franken Riesling. California wineries decided to use Johannisberg Riesling to designate a wine made from the true Riesling grape; then researchers at the University of California at Davis decided that the wine should be called White Riesling. Thus, today in California and elsewhere in the United States the true Riesling grape appears in wines that are labeled either Johannisberg Riesling or White Riesling.

Riesling grapes produce delicate white wines that have delicious aromas and tastes; they are also relatively low in alcohol. Writer William Massee (1970:26) described the Riesling grape as producing "light, dry, flowery wine that has a beautiful balance of freshness and fruit." Long, cool, temperature-controlled fermentation brings out the best in this cultivar. Flowery aromas and fruity flavors, often reminiscent of citrus, peach, or apricot, are common in the best versions of Riesling.

The Riesling grape is also subject to attack by *Botrytis cinerea*—what the French call *pourriture noble*, or noble rot. Botrytised grapes are capable of producing splendid sweet wines, referred to by Hugh Johnson (1994:25) as "the ultimate honeyed, delicate, flower-scented nectar." In the United States, these wines are usually labeled "late harvest," because German terms such as *Auslese* (selected harvest), *Spätlese* (late harvest), and *Trockenbeerenauslese* (made from grapes that have been shriveled by botrytis) are not allowed on American wine labels.

However, Riesling's popularity has waned in recent years, partly because it is made in a variety of styles, most of which are relatively sweet (at a time when American consumers seem to be turning to drier wines) and partly from intense competition from blush wines, especially White Zinfandel. Residual sugar (natural grape sugar that remains unfermented in the finished wine) levels of 2.0 percent or more are still common in most American Rieslings, though the best of these delicate wines are balanced by good acid levels. Some people say these wines can be matched with certain kinds of foods (including Chinese food, some suggest), but to others they are best sipped on their own.

Like Chardonnay, Riesling reaches its optimum quality in the coolest growing climates. Only one county in the Golden State has more than 1,000 acres of this cultivar today—Monterey County, with 1,774 acres. Washington, Oregon, Idaho, Arizona, New Mexico, Texas, Michigan, Illinois, Ohio, Wisconsin, Connecticut, Maryland, New Jersey, Rhode Island, Massachusetts, and New York all have cool niches where Riesling can do well. Without strong consumer demand, however, Riesling acreage has often given way to plantings of Chardonnay.

Cabernet Sauvignon

Cabernet Sauvignon is America's (and many would say the world's) premier red wine grape. It came here from France's Bordeaux region, where it is typically blended with other grapes—including Merlot, Malbec, Petite Verdot, and Cabernet Franc—to produce some of the world's greatest red wines. Cabernet Sauvignon is the major grape in the wines of France's Médoc region, location of the great *châteaux* such as Margaux and Mouton.

Cabernet Sauvignon vines have small berries with tough skins, from which the wines usually acquire a considerable amount of tannin—a group of chemical compounds found in the bark, roots, seeds, and stems of many plants. The wines are often described by wine writers as herbaceous, with hints of green olives and cassis; tinges of mint and sage can be found in some of them as well, and maybe even an occasional touch of black cherry and chocolate. Some develop a black-peppery characteristic, whereas others taste distinctly of bell peppers, though viticultural practices have reduced this particular characteristic considerably. The latter was especially common in some of the earlier Cabernet Sauvignons from the Salinas Valley in California. Cabernet Sauvi-

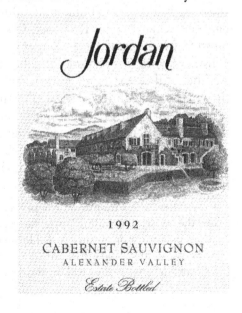

Jordan

1992

CABERNET SAUVIGNON
ALEXANDER VALLEY

Estate Bottled

gnon wines are almost always aged in oak (usually French, sometimes American), which adds layers of flavor and complexity.

Although the Cabernet Sauvignon grape can be grown within a range of climates, it does best where conditions are at least similar to those of France's Bordeaux region. The area around Oakville in California's Napa Valley, for example, is often referred to as the "Bordeaux Belt," an indication of its suitability for Cabernet Sauvignon. Nearby Rutherford is also praised for the quality of its Cabernet Sauvignon, as are many other parts of the Napa Valley. This grape has found some excellent homes in other American viticultural regions as well, including Washington's Yakima Valley (which is producing some splendid examples) and a few locations in Oregon, Arizona, New Mexico, Texas, Missouri, Arkansas, Ohio, Georgia, Virginia, Pennsylvania, Maryland, Connecticut, New Jersey, and New York. As with Chardonnay, the best American Cabernet Sauvignons are indeed world-class wines and can hold their own with the best from Bordeaux.

Pinot Noir

Pinot Noir is the great red cultivar of Burgundy, the grape that provides us with the fine wines of the Côte d'Or; it is also one of two important red grapes in the Champagne region of France (along with Pinot Meunier, which has only recently been planted in small amounts in the United States). In Burgundy we find such royal wine names as Chambertin, Romanée-Conti (among the world's most expensive wines), and Clos Vougeot. Pinot Noir is a grape of well-defined flavors, hence under ideal conditions it leads to great wines, whether still or sparkling. Classic Pinot Noir tastes are often elusive; they may include hints of smoke, leather, mint, plums, and cherries.

The Pinot Noir grape has often been considered difficult to grow, though many winegrowers have been determined to coax the best out of it by finding ideal locales in which to plant and nurture the vines. As chef and wine writer Roy Andries de Groot (1982:136) once noted:

> Everywhere the Pinot Noir has proved itself to be one of the most (if not the most) demanding, difficult, recalcitrant, and temperamental of all grapevines. For even reasonable success, it requires near perfection of growing conditions. Even at its very best, away from Burgundy one seldom finds the joyous feel on the tongue of satin, silk, and velvet.

FINGER LAKES
Pinot Noir
1991

PRODUCED AND BOTTLED BY
HERMANN J. WIEMER VINEYARD, INC., DUNDEE, N.Y.
Alcohol 13.3% By Volume.

Pinot Noir vines are considered weak and delicate; they are also miserly producers, with well-pruned vineyards often yielding less than one ton per acre. They reach optimum quality and flavor in the coolest viticultural regions, and even there, only in selected sites. Within California, Pinot Noir reaches its best in places such as the Santa Cruz Mountains, Los Carneros, and the Russian River Valley. Some wine writers have suggested that Oregon's cool Willamette Valley may become America's premier Pinot Noir region; small acreages are planted in a few other states, including Arizona, New Mexico, Texas, Washington, Idaho, Massachusetts, Rhode Island, Pennsylvania, and New York.

Introducing Some Other Important Cultivars

There certainly is more to the enological world than the noble four alone would allow; a part of wine's appeal is its almost infinite variety. Although the four noble grapes and their geographic distributions certainly deserve the special attention they have already received, they hardly constitute a comprehensive picture of viticulture in America in the 1990s. Many other cultivars are also important, even if their names are sometimes less well known and their wines perhaps a touch less "aristocratic." Because of recent consumer preferences for Chardonnay and Cabernet Sauvignon to the virtual exclusion of some less-appealing varieties (a situation that may now be easing a little, as we shall soon see), acreages of many less-appreciated cultivars are declining; for example, grapes such as Riesling and Gewürztraminer have already experienced serious declines in acreage in some states.

Some grapes, such as French Colombard, Barbera, and Carignane, remain the workhorses of the American wine industry; they are grown in large quantities, primarily in the warmer climate zones of California's Great Central Valley, and are blended into countless vats of generic table wines. Others, including Zinfandel and Sauvignon Blanc, can produce wines that at their best can rival wines made from the noble grapes. However, they are sometimes grown in the warmer climates as well, where their wines may not be as good, though their yields will be high enough to make growing them profitable anyway (remember, the wine industry *is* a business).

It is impossible, of course, to contemplate here the characteristics, utilization, and geographic distributions of all of the wine cultivars that are grown in the United States (the California Agricultural Statistics Service alone currently reports acreage figures for fifty-two different wine cultivars, not counting multiuse grapes such as Thompson Seedless, which can be fermented into wine, eaten fresh, or turned into raisins). However, those that are considered further on are among the most important ones, either in terms of overall acreage or current consumer interest in the marketplace.

Changing consumer preferences are leading winegrowers in some new viticultural directions in the United States, and our discussion of important cultivars begins with three such trends: the rising popularity of many French Rhône cultivars, the renewed interest in Italian cultivars, and the creation of what are now called (thanks to a contest that was held a few years ago to search for a "non-French" name for these blends of Bordeaux grapes!) "Meritage wines."

Despite the fact that the Meritage category, by definition, is composed of blends of wines from different Bordeaux grapes, most Meritage cultivars are also produced as varietal wines. Rhône and Italian cultivars appear both as varietal wines and in Rhône and Italian blends, respectively. In all three categories, some of the included cultivars have been grown in America (mainly in California) for decades (Petite Sirah and Barbera are examples), whereas others are just starting to be planted (Viognier and Nebbiolo, for example).

Cultivars from France's Rhône Valley

Often considered the founding father of the "Rhône Rangers"—a group of California wine makers committed to making Rhône-type wines—Randall Graham (of Bonny Doon) has been a clever and outspoken leader among enthusiasts who began in the 1980s to command more attention for wines made from Rhône grapes, either as varietal wines or as Rhône blends. French Rhône wines underwent a similar rise in popularity at about the same time, partly because of rising prices for the wines of Bordeaux and Burgundy and partly because they were popularized by Robert Parker, perhaps America's most prominent wine critic, in his 1987 book *The Wines of the Rhône Valley and Provence*.

For viticultural purposes, France's Rhône Valley is often divided into northern and southern segments. Syrah and Viognier are the predominant cultivars in the Northern Rhône. The Southern Rhône is best known for its blended wines, including the widely known Châteauneuf-du-Pape (which may legally include up to thirteen different cultivars in its wines). Wines have been produced in the Rhône Valley and transported along the Rhône River for more than 2,000 years. Although it is best known for its red wines, the Rhône Valley produces white wines as well, and at least one white Rhône cultivar (Viognier) has materialized in a few American vineyards recently.

Petite Sirah

Because Petite Sirah was grown in America (almost exclusively in California) about a century before the current popular trend toward Rhône cultivars began, we begin with it, even though most winegrowers omit it from their Rhône blends and it may have no "true" linkage to the Rhône Valley whatsoever. Like Zinfandel, Petite Sirah is a grape that has a somewhat tangled and nearly indecipherable past in California. It was long believed to be the great grape of France's Rhône Valley—the Syrah—and was then thought for a time to have been the lesser-known Durif. But it now appears that California's Petite Sirah is of unknown parentage; it is included here under Rhône cultivars mainly for convenience and because it has been traditionally thought of as a Rhône cultivar.

Petite Sirah produces small black berries in medium-sized clusters; it is believed to be quite resistant to downy mildew. Yields range from

around five tons per acre in the cooler climates to up to eight tons per acre in the warmer climates of the Great Central Valley. However, Petite Sirah is widely distributed in California; Mendocino, Merced, Monterey, Napa, and Sonoma Counties are the leading producers, which demonstrates that the grape is being grown in a variety of different viticultural climates and for a variety of reasons (from being used in jug wines to its use in a few serious, and increasingly expensive, varietal wines). Aside from California, Arizona may have the only other Petite Sirah vines growing in the United States today. By the mid-1990s, Petite Sirah acreage in the United States had declined to just under 2,500, reflecting a combination of the popularity of white and blush wines in the 1970s and 1980s, the dominance of Cabernet Sauvignon and Merlot in the red-wine market, and the renewed interest in Syrah and other "true" Rhône grapes and various Italian cultivars and blends.

Wines made from Petite Sirah are dark in color, which has made it popular as a blending grape in inexpensive red wines; Petite Sirah wines are often described as "inky" and can temporarily stain even the whitest teeth. Characteristically, these wines are quite tannic, especially when young, and they age only slowly. The number of wineries making Petite Sirah as a varietal wine has fallen in recent years, but despite this, prices have been rising.

Syrah

Syrah ranges toward noble status in France (especially in the wines of Hermitage and Côte Rôtie), has become an important cultivar in wines of the Midi region in the south of France (where it is often blended with Cabernet Sauvignon), and has been transplanted successfully as far away as Australia (where it is usually called Shiraz, not Syrah, and reaches its pinnacle in Penfolds's rich and powerful Grange—one of the world's truly great red wines). At its best, Syrah can make wines that rival Cabernet Sauvignon in depth, richness, and aging potential. Considered easier to grow than Cabernet Sauvignon, Syrah is a good producer, resists most common pests and diseases, and is gradually increasing in popularity.

In California, the true Syrah grape of the Rhône Valley is now being produced as a varietal wine by a growing number of wineries. This grape, more elegant and refined than the better-known and more widely planted Petite Sirah, now has a total acreage of nearly 1,000. Although hardly a formidable competitor at this point, interest

in Syrah is growing steadily and its acreage is increasing. At the same time, on the opposite side of the continent, primarily in Virginia, Syrah and other Rhône grapes are finding new homes as well. There is also a small acreage of it in Texas.

Carignane (Carignan)

Carignane originated in Spain (where it is known as Cariñena), though it is most widely grown today in the Midi region of France (where it is called Carignan and has flourished since the eleventh century). It is also grown to a lesser extent in the southern Rhône Valley. Carignane remains the leading grape in France in terms of acreage (though hardly quality) and is the true workhorse of French viticulture. Nevertheless, its acreage has declined considerably in recent years. In America (again, mainly in California), Carignane is something of a latecomer compared to Petite Sirah, having arrived only after Prohibition ended in 1933.

Carignane is a moderate-sized grape, black in color, that grows in medium-sized clusters. Its main appeal is prodigious yields, not high quality; in cooler climates it yields perhaps five to eight tons per acre, though in the warmer regions it may yield up to twelve or thirteen tons per acre. In cool regions it is susceptible to powdery mildew, whereas in the warmest regions it has problems with bunch rot. In the warmest viticultural climates, this grape can maintain reasonable acid and tannin levels, though it can produce more balanced—if neither particularly distinctive nor charming—table wines when it is grown in viticultural climates similar to those suitable for good Cabernet Sauvignon or Zinfandel.

In California, Carignane is planted primarily in the Great Central Valley, where it serves as an important ingredient in bulk wines. In the mid-1990s, there were nearly 9,000 acres of Carignane in the state; San Joaquin and Madera Counties each had more than 2,700 acres, far in excess of the number found in any other single county and reflective of this grape's strong concentration in the state's hottest growing zones. Little of this cultivar is grown elsewhere.

Grenache

Originating in Spain, where it is known as Garnacha (or possibly in Sardinia, according to some ampelographers, where something very much like it is known as Cannonau), Grenache is widely grown in southern France, including the southern Rhône Valley, where it is the most important cultivar in Châteauneuf-du-Pape, for example. It remains Spain's most widely planted red grape and is exceeded in acreage in southern France only by Carignane.

Grenache vines are vigorous, tend toward erect growth, and yield rather light-colored berries that have long been popular for rosés because they lack sufficient color to make darker red wines without blending them with other grapes such as Syrah. Grenache does well in most viticultural climates; however, yields are higher and more consistent in the warmer regions. There were more than 12,000 acres of Grenache in California in the mid-1990s, concentrated primarily in Fresno, Kern, and Madera Counties. An increase in Grenache plantings has occurred in recent years, reflecting the growing interest in Rhône-style wines in America; so far, however, little of it is grown outside California.

Mourvèdre (Mataro)

Mourvèdre is another red grape from Spain (where it is known as Monastrell), but it has long been established in parts of the southern Rhône Valley and is important in many Rhône blends. Mourvèdre is relatively low in acidity, often lacks deep color, and yields less than Carignane, though it is also less subject to powdery mildew. Mourvèdre vines are reasonably vigorous, have a good upright growth pattern, and do best in the warmer viticultural climates. In the mid-1990s, there were only 335 acres of Mourvèdre in California (where it is known as Mataro and has begun to claim high prices), two-thirds of which were planted in Contra Costa

County. Virginia also has a few acres planted in Mourvèdre, near Charlottesville.

Viognier

Viognier is the great white grape of the northern Rhône Valley; it ascends to its enological apogee in the wines of Condrieu, known for their distinctiveness (and price!). As British wine writer Jancis Robinson (1986:180) noted:

> Quantitatively, the Viognier vine hardly deserves a mention in this book. Any qualitative assessment of the world's wines, on the other hand, tends to linger over the intriguing Viognier. . . . Viognier produces full-bodied, golden wines with a haunting and tantalizingly elusive bouquet. Some say may blossom, others apricots; others musky peaches . . . even ripe pears just below the skin. . . . Viognier works a magic of its own.

Although reasonably drought-tolerant, Viognier vines are often troubled by powdery mildew. They are also miserly producers, yielding mostly light crops of those dark yellow grapes that give Condrieu its characteristic color. In recent years, Viognier has been planted in the south of France, as well as in scattered parts of California and Virginia, but there are still fewer than 300 acres of this cultivar in America in the mid-1990s. Whether Viognier will produce wines comparable to Condrieu in these new locations remains to be seen, though a few California producers and at least one Virginia winery are already offering varietal versions. Low-priced Viognier (far from Condrieuesque) from the south of France is now arriving in American markets as well, further clouding the Viognier outlook in the United States.

Italian Cultivars

Italy has offered much to enrich American culture, from fine movies (e.g., *La Dolce Vita*) and music to gastronomy—just witness the current popularity of Italian food (isn't there at least one new pasta place near you?) and wine (still harder to find than pasta in most neighborhoods). American wine makers have recently awakened to new viticultural and enological possibilities offered by Italian vines.

Although many Italian cultivars, including Barbera, have been grown in California for decades and some Barbera vines can even be found in New Mexico and Texas, there has recently been a trend toward using cultivars such as Nebbiolo (from Piedmont) and Sangiovese (from Tuscany), both of which are at the apex of wine quality in Italy. Renewed interest in Italian cultivars so far is a mixed blessing; it is good to see new red wines appearing on the market, but stiff prices are likely to slow long-term consumer acceptance, especially while many very good Italian Chiantis are available for much less in the United States.

Barbera

Barbera arrived early in California from the Piedmont region of Italy, where it has long been grown around such towns as Asti and Alba. Asti, of course, gave its name to the tiny town in Sonoma County where the original Italian Swiss Colony winery was founded (and where wines under that label were made for decades). Good Barberas—a mainstay for wine drinkers in northern Italy—are usually tart and fruity, with at least a medium red color. The best California Barberas are typically made from blends of coastal grapes and grapes grown in the Great Central Valley.

High natural acidity and substantial yields make Barbera attractive to grow in the warmest viticultural climates, where most plantings have occurred in recent years. In the mid-1990s, there were more than

EBERLE

1993

PASO ROBLES

BARBERA

NORMAN VINEYARD

PRODUCED AND BOTTLED BY
EBERLE WINERY, PASO ROBLES, CALIFORNIA
ALCOHOL 13.6% BY VOLUME

11,000 acres of Barbera in California, most of which were growing in the Great Central Valley; acreages in New Mexico and Texas are younger and much smaller. The Wines made from this grape are used in the United States mainly for blending into generic burgundies and inexpensive proprietary wines, though a few wineries offer Barbera as a varietal wine. Often associated with spaghetti and other pasta dishes (now fashionable as well, especially because of the rising popularity of the so-called Mediterranean Diet, with its emphasis on complex carbohydrates such as pasta and rice, fresh fruits and vegetables, olive oil, and, of course, wine), it is equally good with a host of other well-seasoned foods. Both Mediterranean and California cuisine would find Barbera an excellent companion, if only more producers would take it more seriously and offer it at a reasonable price.

Nebbiolo

Nebbiolo is the great red grape of Italy's Piedmont (a wine-producing region on the southern foothills of the Alps, with a concentration of Nebbiolo production around Alba and neighboring villages). It is the enological luminary in Barolo and Barbaresco, the grape of lesser-known but well-loved wines such as Spanna (a colloquial name for the grape in parts of Piedmont), Gattinara, and Ghemme. Jancis Robinson (1986:147) wrote about Barolo and Barbaresco that "so revered are these wines throughout the major population centres of Italy that they were the natural first candidates for elevation to the newly created rank of DOCG [Denominazione di Origine Controllata e Garantita]." If Nebbiolo had been more widely known, then it would possibly have ranked among the noble grapes, but the Italians have long hoarded it. The best Barolos rank among the world's greatest red wines, tinged with aromas that run from tar to violets and truffles; they can age for decades. In America, Nebbiolo is a newcomer, a welcome addition to the red-grape roster. Whether American Nebbiolos can rival those of Piedmont remains to be seen, however. These wines can be highly tannic and acidic and need many years to fully mature (at a time when most American wine makers are in a rush to market recent vintages for cash flow purposes and when consumers are eager to drink what they have just purchased).

Nebbiolo produces only moderate yields, and at one time it was recommended for planting in the warmer viticultural climates in the United States; it may do even better in those same areas in which

Cabernet Sauvignon does well. Nebbiolo grows vigorously and is resistant to most parasitic infestations; oidium, however, can be a problem. In the mid-1990s, California's Nebbiolo acreage just reached 100, up from fewer than ten acres a decade earlier; other states have not yet done much at all with this cultivar.

Sangiovese (Sangioveto)

Sangiovese (officially listed in California as Sangioveto) is the leading red grape used in Chianti, probably Italy's best-known wine, long associated with the basket-covered bottles (*fiaschi*) in which it often came (and which have largely disappeared as the quality of Chianti has substantially improved). Sangiovese and Barbera are the most widely planted grapes in Italy.

Sangiovese has been recommended for planting in relatively warm viticultural climates in the United States, though it may also do well in cooler areas. This cultivar is moderate to high in yield, resistant to most grape infections, and moderately vigorous in growth habit. As of the mid-1990s, there were over 600 acres of Sangiovese in the Golden State; interestingly, the largest plantings are in Napa and Sonoma Counties, not in the warmer climes of the Great Central Valley. Other states have so far shown little or no interest in this cultivar.

Meritage Cultivars

France remains the major source of other cultivars for the American wine industry, in addition to contributing three of the four noble grapes—Chardonnay, Pinot Noir, and Cabernet Sauvignon. France also remains the viticultural and enological standard, not just for America but for virtually all of the world's wine-producing regions. As Hugh Johnson (1994:52) commented, "When the last raindrop has been counted, and no geological stone is left unturned, there will still remain the imponderable question of national character which makes France the undisputed mistress of the vine."

Meritage blends, both white (mainly Sauvignon Blanc and Sémillon) and red (a combination of some or all of the following: Cabernet Sauvignon, Merlot, Cabernet Franc, Petite Verdot, and Malbec), use Bordeaux grapes almost exclusively; they have become increasingly popular and represent some of the finest wines that America currently has to offer. Wineries must pay a fee to use the name Meritage on wine labels, so many smaller wineries make such blends but do not refer to them as such.

Sauvignon Blanc

Like Chardonnay, Sauvignon Blanc is another white grape whose ancestral home is in France, albeit from a different viticultural region. It is at its best in many Bordeaux white wines (often blended with Sémillon), especially those of Graves, and it also appears in the Loire Valley in the fashionable wines of Sancerre (which especially appealed to Ernest Hemingway) and Pouilly-Fumé. Sauvignon Blanc, also called Fumé Blanc by some American vintners, has been grown in the United States for at least one hundred years. During that time, it has had its ups and downs, its cycles of cheerful popularity and relative obscurity. However, since the 1980s the Sauvignon Blanc grape has definitely been riding the crest of one of the high points in its popularity cycle. Sometimes referred to as a "poor person's Chardonnay," Sauvignon Blanc in the hands of wine makers such as California's Fred Brander and Jon Bolta can be made into excellent white wines.

Sauvignon Blanc grapes are typically golden yellow in color and grow in small clusters. This cultivar yields reasonably well and has been planted throughout most of the cooler viticultural regions of California, as well as in Oregon, Washington, Idaho, Arizona, New Mexico, Texas, Georgia, North Carolina, Virginia, Maryland, Pennsylvania, New Jersey, and New York. There are now perhaps 3,000 acres of Sauvignon Blanc growing on American soil.

One of the first importations of Sauvignon Blanc to California came from the famous Château d'Yquem, which is world-renowned for its sweet botrytised wines. Those early vines were planted in the Livermore Valley, which was later to become well known for its white wines. Ideal for this grape are locations where conditions are similar

GRGICH HILLS

Napa Valley
FUMÉ BLANC
Dry Sauvignon Blanc
1993

PRODUCED AND BOTTLED BY GRGICH HILLS CELLAR
RUTHERFORD, CA · ALC. 12.8% BY VOL. · CONTAINS SULFITES

to those in parts of Bordeaux and the Loire Valley in France. Livermore is relatively warm, but this grape indeed continues to thrive there despite growing threats from the urbanization that spreads ever outward from the San Francisco Bay Area.

American Sauvignon Blancs are distinctive, especially when grown in the cooler regions, where they take on grassy-herbaceous notes that can be quite extraordinary. The best of these wines are treated much like Chardonnays at the winery, receiving oak aging (sometimes even fermentation in oak) and some bottle aging. Wine writer Anthony Dias Blue (1982b:22) made the following comment about Sauvignon Blanc: "This noble wine does have a flavor and aroma that, at its best, smells like new-mown hay. The finest examples also have elegance and breeding. They are second only to Chardonnay in complexity and finesse, and they have a raw-boned assertiveness that is all their own."

Roy Andries de Groot (1982:56) wrote (quite correctly, it turns out), "I firmly believe that the greatest days of the Sauvignon Blanc in America are still in the future—with magnificent Sauvignon Blanc tasting experiences to come for all of us." Those who are attracted to this grape and its fine wines are most assuredly glad that he was right. Among American consumers, however, preference has been shifting away from the grassy style toward wine with more fruit flavors and hints of vanilla from oak.

Sémillon

Like Sauvignon Blanc, Sémillon is another white grape that came to America from Bordeaux, where it is used both for blending with Sauvignon Blanc in many white Bordeaux wines and for producing rich, sweet, botrytised dessert wines. Sémillon is a good-yielding grape, capable of producing five to ten tons per acre, depending on local environmental conditions and pruning techniques. Although vigorous in growth habit, Sémillon is susceptible to rot, including the beneficial botrytis; the vine's golden yellow, distinctively flavored grapes grow in small clusters. Nearly 1,700 acres of Sémillon were growing in the Golden State in the mid-1990s; there are also small acreages in Washington (where several wineries have produced it, very successfully, as a varietal wine), Idaho, Texas, and Virginia.

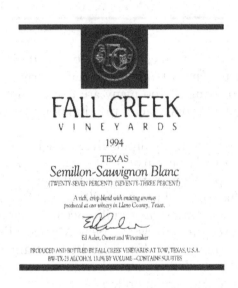

Merlot

Merlot, a leading red grape in Bordeaux (especially prominent in Pomerol) and a major grape for blending with Cabernet Sauvignon both there and in America's red Meritage wines, has achieved considerable popularity on its own in America as a varietal wine. In the mid-1990s, scarcity is driving prices up for Merlot, especially with the

renewed demand for red wines, and quality is suffering a bit as a result. As wine writer James Laube (1996b:35) warned recently about current Merlot offerings, "Too many are marked by simple, earthy, swampy and vegetal flavors, void of the kind of pure fruit essence that lures most wine lovers."

Merlot vines are reasonably productive, moderate in growth habit, produce medium-to-large clusters of medium-sized bluish-black berries, and have problems with downy mildew. By the mid-1990s, Merlot acreage had reached nearly 18,000 in California alone, and it had also become of at least some significance in Oregon, Washington (where it does exceptionally well), New Mexico, Texas, Georgia, North Carolina, Tennessee, Virginia, Pennsylvania, New Jersey, Connecticut, and New York. It grows best in the same viticultural climates that Cabernet Sauvignon prefers.

Cabernet Franc

Less well known than Merlot, Cabernet Franc is another red grape from Bordeaux, where it has been used as a blending grape with Cabernet Sauvignon and other cultivars. It is also grown in Italy, where it is sometimes known as Cabernet Frank. Some ampelographers suggest that given its overall similarities to Cabernet Sauvignon,

Cabernet Franc may actually be a mutation, better adapted to some localized set of environmental conditions in cooler parts of Bordeaux.

Cabernet Franc tends to produce wines that are somewhat lighter and less tannic than those made from Cabernet Sauvignon, so its role in a blend is similar to that of Merlot, used to soften the harsher tones created by Cabernet Sauvignon. Cabernet Franc is more vigorous and yields somewhat better than Cabernet Sauvignon; it can have problems with mildew, however.

As Jancis Robinson (1986:134) perceptively noted, "Although Merlot is a more obliging producer, it seems likely that plantings of Cabernet Franc will increase perceptibly in California as the need for a suitable complement to Cabernet Sauvignon in top quality wines is recognized." She could have said that of other states as well. Between 1986 and 1994, Cabernet Franc acreage in the Golden State increased from 760 to just over 1,900 (still little more than one-tenth of the Merlot acreage, however). Washington, Georgia, Virginia, Maryland, Connecticut, Pennsylvania, and New York are also growing some Cabernet Franc, mainly for blending with Cabernet Sauvignon. Those who wish to continue consuming wines with meals without mortgaging their homes to buy them should hope for more plantings of cultivars such as this one.

Petite Verdot (Petit Verdot)

In recent years Petite Verdot, still another red grape grown in Bordeaux (where it is known as Petit Verdot), has commanded the highest price of any California wine grape in some recent years (an average of $1,599 per ton in 1994, compared to $862 per ton for Cabernet Sauvignon). Its high price reflects both an increased appreciation of

ALEXANDER VALLEY
PETIT VERDOT
1991
ALC. 12.8% BY VOL.

its high quality and its current scarcity. Dark in color, Petite Verdot is capable of producing deep-hued wines that can add depth of color and variety of flavor to Meritage blends.

Like Cabernet Sauvignon, Petite Verdot is fairly resistant to rot; its late ripening habit, however, can be problematic, especially in areas subject to fall rains or frosts. In 1994, there were only 160 acres of Petite Verdot in California, up from a mere 20 acres in 1986; some very small acreages may be currently planted elsewhere in the country.

Malbec

For the sake of completeness in our discussion of Meritage blends, Malbec, another red grape from Bordeaux (where it is also known as Cot) and Cahors (where it is also known as either Cot or Auxerrois), deserves mention, even though it has the least acreage of all the Meritage cultivars in the United States. According to Jancis Robinson (1986:198), Malbec "produces a sort of watered-down rustic version of Merlot, mouthfilling and vaguely reminiscent of blackberries in youth but soft and fairly low in acid, and, therefore, early maturing."

In Bordeaux, Malbec acreage has been declining for many years; Merlot is preferred to Malbec by most current Bordeaux wine makers. In California, Malbec acreage grew from 25 in 1986 to 98 in 1994; it remains to be seen whether it is likely to increase in importance in this country, especially with the widespread planting of Merlot and the rapidly expanding acreage of Cabernet Franc.

Other Important Cultivars in American Vineyards

Chenin Blanc

The importance of Chenin Blanc is apparent when we look at the acreage for this cultivar. However, before that we should glance at the grape itself and the variety of wines that can be made from it. Also, we should note that sales of Chenin Blanc (along with Riesling and Gewürztraminer) have suffered in the United States as a result of competition from blush wines, especially White Zinfandel. However, some wine makers have been taking a new look at this grape and making drier wines from it—and some are very good.

Chenin Blanc came to America from the Loire Valley in France, where it has thrived comfortably for perhaps one thousand years and

True Frogs

SIERRA FOOTHILLS

CHENIN BLANC

CELLARED AND BOTTLED BY RIBBET CELLARS
MURPHYS • CALAVERAS COUNTY • CALIFORNIA

ALC. 12% BY VOL • CONTAINS SULFITES

is sometimes known as Pineau de la Loire; it is the grape of Vouvray and many lesser-known wines of the Loire. In France, wines made from this grape are fashioned in a myriad of styles, from dry to sweet, from still to sparkling. In the United States, the wines are made in a variety of styles also, but the overwhelming favorite seems to be slightly sweet; the best of them have sufficient acidity to give the wines a nice balance. These wines are popular and are an excellent introduction to varietal wines for people who have just been ordering their favorite American generic white wine. Chenin Blanc is fashionable in South Africa, as well, where it is known as Steen.

In the mid-1990s, there were nearly 30,000 acres of Chenin Blanc growing in the United States. In cooler climates the grape is used to make varietal wines like those described above. In the warmer climates, this grape's tendency toward high natural acidity helps it produce pleasant and drinkable table wines, despite the often intense summer heat of its major American home, California's Great Central

Valley. Chenin Blanc can also be found in Washington, Idaho, Arizona, New Mexico, Texas, Georgia, and Virginia.

In an article ascribing various personalities to different wine grapes, wine writer Anthony Dias Blue (1983:18) wrote: "Finally, there is Chenin Blanc—quiet, unobtrusive, good-natured Chenin Blanc, the workhorse of the family. Chenin is plainer than her flashier sisters, but she has a heart of gold. Fresh-faced, youthful, willing and adaptable, she can be either soft and sweet like Riesling, or crisp and angular like Sauvignon Blanc."

French Colombard

French Colombard came to America (almost exclusively to California) late in the nineteenth century from the Charente region of France, where it is used primarily to make wines that are then distilled into that famous brandy known as Cognac. It is a high-yielding grape that can average seven tons or more per acre in California's hot Great Central Valley; it may produce nearly twice that if conditions are ideal. Even in the cooler viticultural regions, it can yield five or more tons per acre. With its excellent acidity and high yields, it has traditionally been grown in the warmer climate regions and has been used as a blending grape in generic Chablis and Sauterne. However, since the 1960s it has also been used to make a few varietal wines, the best of which are crisp, flavorful, and show an excellent balance between a touch of residual sugar and the typical high acidity.

With nearly 50,000 acres planted in the mid-1990s, French Colombard is second in acreage to Chardonnay among California's wine grapes. The most recent plantings of French Colombard have been in the Great Central Valley, though it has been planted in smaller amounts in many of the cooler coastal regions as well; other states with some French Colombard acreage include Arizona, New Mexico, and Texas.

Gewürztraminer

Gewürztraminer, thought of as a German grape even though it grows in France's Alsace-Lorraine region as well, seems to suffer from a variety of problems in the United States, not the least of which is the difficulty of pronouncing its rather unwieldy Germanic name. Despite its apparent tongue-twisting qualities, the word can be subdi-

vided and easily understood. Traminer (pronounced "trah-minn-ur") is the name of a grape, and gewürz (pronounced "geh-virts") means spicy, so when we put them together we have the spicy Traminer grape. Despite its German name, the grape is perhaps better known in America for its Alsatian wines. However, American wine makers have tended to make Gewürztraminer in a more Germanic style, though this tendency has diminished somewhat in recent years.

Gewürztraminer grows in small clusters, with berries that are at least tinged with pink. Occasionally, there is sufficient color to produce a delicate rosé, though very few such wines are produced in this country. Gewürztraminer grows best in the coolest viticultural climates, where it can produce exceptionally crisp and spirited wines with an excellent balance between acidity and a slight amount of residual sugar. There are perhaps 2,000 acres of this cultivar in America, where it is grown in small acreages from New York, Rhode Island, Connecticut, New Jersey, Pennsylvania, and Ohio, through Texas and Colorado, to Washington, Oregon, and California.

In addition to its difficult name, Gewürztraminer has other problems—it can be a viticultural nightmare. First, it is difficult to grow and must be harvested at the precise moment of its full ripeness in order to yield its best wines. Second, this grape has a characteristic bitterness in the aftertaste that is imparted to the wine, so that most producers are tempted to leave a bit of residual sugar in their wines in order to mask it. Third, the wines are often intensely flavored, so people may tire rather quickly of drinking them. It is a prominent wine, however, used for demonstrating to people just how distinctive a white varietal wine can be, especially for people who are relative newcomers to wine and who may feel that all white wines are quite similar in their aromatic characteristics. It also is recommended by numerous wine writers as a particularly good accompaniment to a variety of Asian foods.

Zinfandel—California's Unique Cultivar

The Zinfandel grape has been, at least until recently, unique to California; it is also the only cultivar to have its own fan club, ZAP (Zinfandel Advocates and Producers). Although its parentage is distinctly European *Vitis vinifera*, it is a grape whose ancestry has been the source of considerable debate and mystique, a viticultural orphan left to fend for itself. Stories that have tried to explain the heritage of this grape have often only further shrouded it in mystery. One story is that Zinfandel was a mysterious grape brought to California by Agoston Haraszthy, one of the state's most colorful wine figures, who planned to plant it in his vineyard near the town of Sonoma, where we now find the Buena Vista Winery. As legend went, labels were lost or misplaced in shipping and the heritage of this grape was forever lost. Furthermore, no similar cultivar could be found in modern Hungary.

This was indeed a good story, but as ampelographers began to investigate it, the romance of the story yielded to a somewhat less mysterious reality. Professor Olmo, then at the University of California at Davis, argued that Zinfandel could be traced back to an Italian heritage, to a grape called Primitivo di Gioia. In Italy this grape grows in Apulia and makes wines that are not at all renowned in that country.

Subsequent to the work of Professor Olmo, others searched out fragments of the story as well, especially those dealing with the arrival of Zinfandel from Europe. Roy Andries de Groot (1982:160) found that "there are in existence unquestionable documentary records of a Black Zinfandal Vineyard on Long Island, New York, in 1830, and another, in the 1840's, at Salem, Massachusetts." The 1830 nursery catalog of William Prince of Long Island listed a Black Zinfandel. The name, though not the vine, may have been a corruption of the Austrian Zierfandler. In any case, regardless of its diffusion path, Pro-

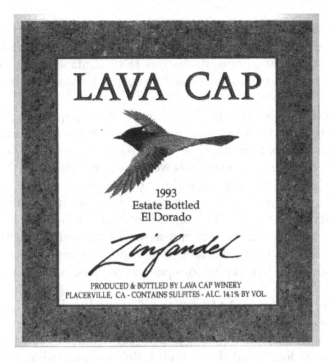

fessor Olmo was, until quite recently, believed to be correct in suggesting that Zinfandel, as we know it today, derived from the Primitivo variety mentioned above. It now is believed, however, that the Primitivo arrived in Italy some time *after* it arrived in America.

For some time, a few ampelographers believed that Zinfandel might also share some link with the Plavac Mali grape found in Dalmatia. In confirming that view, and perhaps finally ending the search for Zinfandel's roots, wine writer Terry Robards (1996b:46) recently wrote that "an array of newfound evidence indicates that Zinfandel came from what is now Croatia in Eastern Europe, where its relative is variously called Plavac Mali or Mali Plavac and is used to make Croatia's greatest wines, among them Dingac and Postup from the Dalmatian coast on the Adriatic Sea." Robards also noted that this region was once part of the Austro-Hungarian Empire, from which Haraszthy returned with his cuttings in 1861, adding credence to the original story that Haraszthy had brought this cultivar to California. Although the mystery of Zinfandel's heritage has now seemingly been resolved, there remains the mystery of how it got its name.

It also seems clear that Zinfandel was first planted in California in the decade following the gold rush, and it certainly found the Golden State well suited as a home. As early as 1883, Zinfandel appeared as a varietal name on California wine labels, and its wines were said to have been well received.

Zinfandel is a reddish-black to almost black grape that grows in medium to large clusters; uneven ripening of berries within clusters sometimes creates a problem for Zinfandel producers. The Zinfandel is a thin-skinned grape that yields well in warmer climates, though it is widely grown throughout all but the coolest coastal regions. It seems at its best for good table wines in the more moderate viticultural climates. In the warmer zones, it produces wines that are more raisiny than berryish and are not as well balanced, though in the warmer climates it can produce very good port-type wines. Yields range from about five to nine tons per acre, depending on climate and soil conditions.

In 1960, there were 23,618 acres of Zinfandel grapes in California, and in 1994, there were almost 39,000 acres, with the greatest concentration (more than 25,000 acres) in San Joaquin County; Zinfandel is widely distributed throughout California's viticultural regions, though its concentration in the delta region of the Great Central Valley is hard to overlook. It has also been transplanted to Oregon, New Mexico, Texas, and Virginia in recent years.

Zinfandel has served a variety of vinous functions. As Anthony Dias Blue (1981:12) pointed out, "For years, carloads of good, unpretentious red table wines were produced using mostly Zinfandel grapes, and even today much of California's red jug wine . . . is blended with Zinfandel." However, Zinfandel is capable of rising above this humble but important role—at its best it can even rise to greatness. Since the mid-1970s, it has filled the role of blush wine almost too well, with White Zinfandels being made now by numerous wineries. Zinfandel has often posed a problem for consumers because of its many guises—it has been made in a variety of styles, from white to "monster" red, "late harvest," and even "port." In the best of the red Zinfandel table wines, we find a delicious berryish taste, generally described as more like raspberries than any other, though blackberries are mentioned in descriptions also, as is "briery." As with other table wines, the best Zinfandels are produced from grapes that have been grown in the cooler regions, where the grapes can achieve the best balance between acid and sugar levels. During the 1970s, Zinfandels from old vines growing in the foothills of California's Sierra Nevada, primarily in Amador County, became quite popular. As Roy Andries de Groot (1982:163) noted some time ago, "Clearly, Zinfandel, in terms of prestige and quality, has moved from being an everyday working wine to becoming one of the expensive and major labels of California." Indeed, Zinfandel can be good, even great, but the consumers often have a hard time finding their way around the many styles because the label seldom gives a clue to the contents.

Zinfandel lost favor in the 1970s and 1980s, just as most other red wines did. Only the rising popularity of White Zinfandels kept many of these vines from being torn up and replanted or grafted over with some popular white cultivars. Since 1991, however, the growing popularity of red wines as a "health food" for the heart has increased the demand for red Zinfandels again. Good Zinfandels can compete favorably for the red wine market, and we can hope that the taste for red wines remains strong. Many wineries have taken a genuine interest in making quality Zinfandel, and California viticultural regions such as the Sierra Nevada foothills and Dry Creek Valley have become especially well known for high-quality Zinfandels.

Important Non-*Vinifera* Cultivars

Those accustomed to talking about California wines alone among American wines are likely to encounter some unfamiliar cultivars if they begin looking at vineyards and their wines in the eastern United States. Both native American and French-American cultivars continue to play major roles in most eastern vineyards.

A Few Native American Vines

As noted already, wine growing in the eastern United States was confounded by the early unsuccessful introduction of *Vitis vinifera*; the first enological successes came with grapes from native American vines, primarily *Vitis labrusca* and American hybrids, some of which remain important today. Although most of these natives are resistant to fungal infections and phylloxera, many are especially susceptible to black rot and Pierce's disease and can have problems with powdery mildew as well. Wines from these grapes have very different characteristics than those made from *Vitis vinifera* cultivars; their aromas and tastes are often characterized as "foxy," a term described by Jancis Robinson (1986:228) as "oozing the musky smell of a wet and rather cheap fur coat." These grapes are today best known for their ubiquitous use in juices and jellies, not their use in wines.

Concord

First planted in 1843 by Ephraim W. Bull in Concord, Massachusetts (where it got its name), this is a classic *Vitis labrusca* grape. High yields (often ten tons per acre), good resistance to most vine diseases, and winter hardiness have endeared this cultivar to eastern viticulturists for more than 150 years. It remains the most widely planted grape east of the Rocky Mountains and is also the leading cultivar in Washington state (though it is not used there for wine). Its primary uses, however, remain in juice and jelly, though wine makers still ferment it in a number of states, often blending it with other grapes into red wines. Although the largest acreages of Concord are found in Washington and New York, it is also grown commercially in West Virginia, Pennsylvania, Ohio, Illinois, Michigan, Missouri, Arkansas, Tennessee, and Alabama.

Catawba

Even older than Concord and second only to Concord in its importance in viticulture in the eastern United States, Catawba is another hybrid of *Vitis labrusca*, perhaps crossed with another American species. Geographer John Baxevanis (1992) has suggested that one of its parents may actually have been a *vinifera* cultivar. It is another winter-hardy and disease-resistant cultivar and is considered by most enthusiasts to be better for wine making than Concord; its wines range in color through a series of shades of pink.

Catawba has been especially popular for use in eastern sparkling wines. This cultivar is most important in the Finger Lakes region of New York and grows well along the south shore of Lake Erie as well, a region that includes segments of New York, Pennsylvania, and Ohio. Additional acreages of Catawba are found in West Virginia, Kansas, Missouri, and Alabama.

Niagara

Niagara is a white American cultivar, thought to be a cross between Concord and Cassady. It is somewhat more susceptible to various grape diseases than either Concord or Catawba and is less winter hardy as well. Nonetheless, it does well in many eastern locations and is the leading white grape east of the Rocky Mountains. Wines made from pure Niagara are extremely foxy, so most wine makers blend in juice from other cultivars, often from French-American hybrids, to neutralize its strong flavor. Niagara appears in vineyards in New York, Pennsylvania, West Virginia, Virginia, Michigan, Ohio, Illinois, Arkansas, Alabama, and Oregon.

Some Popular French-American Hybrids

Most of these grapes have been developed by French hybridizers (including Messieurs François Baco, Victor Villard, Jean-Louis Vidal, Bertille Seyve, and Albert Seibel—names still seen on many such cultivars), though their use in France has diminished considerably since about 1950. Since the 1930s, a number of these cultivars have played an important role in eastern and midwestern viticulture in the United States. They are more prone to diseases that attack *Vitis vinifera* than are the native Americans and are less winter hardy as well. However, wines produced from them are more acceptable to many wine consumers than those from native American grapes—they minimize or even lack traces of foxiness.

Seyval

Seyval is the leading white French-American hybrid cultivar in the eastern United States. It has large, compact clusters of grapes, ripens about midseason, is relatively winter hardy, and produces high yields on most types of soils. Outside the United States, Seyval has done well in both northern France and England, where there has been a resurgence of wine production in recent decades. Wines produced from Seyval, however, lack distinctiveness, though they also fortunately lack any noticeable trace of foxiness. In America, Seyval is widely grown east of the Rocky Mountains and can be found in New York, Pennsylvania, Massachusetts, Rhode Island, Maryland, Virginia, West Virginia, Indiana, Illinois, Ohio, Michigan, Wisconsin, Minnesota, Missouri, Arkansas, Tennessee, Georgia, and Mississippi.

Vidal Blanc

Vidal Blanc, a cross between Ugni Blanc and Rayon d'Or, is another popular white French-American hybrid in the eastern United States. This cultivar can attain high sugar levels in cool growing environments, has reasonably high acid levels, and produces wines that are rather neutral in flavor, with a light touch of fruitiness. The vine has good resistance to most diseases, produces high yields, and grows easily; these features, rather than the character of its wines, account for its popularity with growers. Like Seyval, Vidal Blanc is widely planted in the eastern United States. It can be found in New York, Pennsylvania, New Jersey, Rhode Island, Maryland, Massachusetts, West Virginia, Virginia, South Carolina, Georgia, Arkansas, Missouri, Tennessee, Indiana, Ohio, Michigan, Minnesota, and even New Mexico.

Vignoles (Ravat)

Vignoles, another white grape, has become more popular because it has high sugar and acid levels that allow it to produce exceptional sweet wines, though its yields are lower than those of many other French-American and American hybrids. Its flavor is quite distinctive but not foxy; drier versions tend toward undesirable bitterness. Vignoles is especially important in New York, Pennsylvania, Ohio, and Michigan; however, it is also grown in small amounts in Connecticut, Maryland, Illinois, Wisconsin, Missouri, Arkansas, and Tennessee.

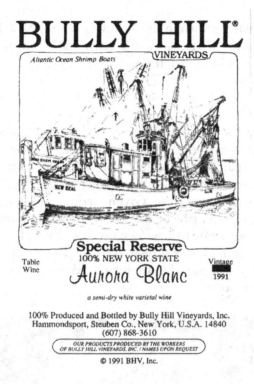

Aurore (Aurora)

Spicy and flavorful, wines made from the white Aurore have made that grape another important cultivar in the eastern United States; it is used in both table and sparkling wines. The vine yields well, has good winter hardiness, and is susceptible mainly to black rot. It is now growing in New York, New Jersey, Pennsylvania, West Virginia, Michigan, Ohio, Indiana, Missouri, Arkansas, and Texas.

Chambourcin

One of the few French-American hybrids to increase in acreage in France in recent decades (especially in the Loire Valley), Chambourcin is a vigorous and high-yielding vine with good disease resistance. It has become one of the most important red French-American hybrids in the eastern United States; its wines can be quite good, tend to be deeply colored, and can benefit from aging in wooden barrels. In some locations, late ripening can be a problem, however. Chambourcin is currently grown in the following states: New Jersey, Pennsylvania, Maryland, Virginia, South Carolina, Tennessee, Arkansas, Missouri, Illinois, Ohio, and Texas.

Chancellor

Chancellor, another red grape, was once the most widely planted French-American hybrid in France, though it has now virtually disappeared from that nation's viticultural landscape. Although Chancellor yields reasonably well, it is not very winter hardy and is susceptible to powdery mildew. Given proper attention, however, wines from this grape can be among the best reds from such hybrids. This grape seems to have done especially well in New York's Hudson Valley, though it is grown in other places as well, including Connecticut, New Jersey, Pennsylvania, Michigan, West Virginia, Georgia, and Missouri.

Baco Noir

Baco Noir is still another important red grape in eastern American viticulture, though it may be declining in popularity somewhat. It is a cross between an unknown variety of *Vitis riparia* and Folle Blanche (a popular grape in France's Loire Valley). Baco Noir ripens early, giving it some advantage in cooler growing zones. It is only moderately winter hardy, but it is a vigorous vine with good disease resistance and its grapes produce well-colored red wines. Geographically, its distribution is somewhat limited compared to other American and French-American hybrids. Currently, it is grown in New York, New Jersey, West Virginia, Ohio, and Illinois.

Summary

Grapes are the fundamental characters upon which the drama of wine production depends. Good wines can only be made from good grapes, which must in turn come from healthy vines that are growing in or near their optimum environments. *Vitis vinifera* cultivars remain the premier vines in American viticultural landscapes and almost exclusively dominate the wine industries of the western states. Although *Vitis vinifera* cultivars have increasingly found niches in states east of the Rocky Mountains as well, eastern American viticulture still depends heavily on American and French-American hybrids to round out the cast.

Because of the need for winegrowers to find the best natural environments within which to grow different cultivars, we turn next to a consideration of the physical environment and vines. Mesoclimates, soils, natural hazards, and pests constrain viticulture within selected regions. These regions, in turn, provide homes for some very different collections of grapevines, which affects the consequent viticultural landscapes to a considerable degree.

3

*A*merican Environments for Wine Grapes

As CALIFORNIA WINEGROWER Justin Meyer (1989) succinctly (albeit modestly) put it, "Quality starts in the vineyard. . . . Given high-quality fruit, it is difficult for even an amateur to foul up what nature has provided." Conversely, without high-quality grapes even the best of wine makers cannot hope to turn out a splendid wine. Cultivars are likely to be healthiest and produce their best grapes when they are grown in or near their environmentally optimum location— where they have the best possible combination of climate, soil, and other natural conditions to ripen them to a peak of perfection.

Of course, some cultivars are more finicky than others, more sensitive to subtle environmental differences. A few can be pushed toward their environmental margins, especially if economics favors such growth, whereas others cannot stray far from optimal conditions without suffering considerable qualitative declines. All things being equal, the best physical environments for grapes are likely to be occupied by cultivars that produce the best wines and the best financial returns to growers. Cool viticultural climates in the United States today increasingly support Chardonnay and Pinot Noir at the expense of Riesling and Gewürztraminer, for example, because the market for the latter two is weaker, keeping their prices relatively low. Grapes such as Barbera, by contrast, are pushed toward the hotter, more marginal viticultural climates, even though wines made from these grapes in cooler growing areas is likely to be of better quality, because Barbera wines are relatively inexpensive compared to Cabernet Sauvignon, Zinfandel, or Meritage blends. Thus, as we consider environmental conditions for growing grapes across America, keep in mind that the combination of cultivars found in a particular viticultural region is a reflection of both that region's environmental characteristics and the prevailing market conditions for American wines.

Given modern wine-making techniques, it is possible to make sound, satisfying everyday wines from a multitude of grape varieties that have been grown under various conditions, so long as the grapes have been able to reach a reasonable degree of maturity, as measured by their sugar content. However, the best wines can be produced only when conditions are near ideal for bringing a specific grape variety, or perhaps more than one variety, to a peak of perfection in which there is harmony and balance within it—that is, sufficient ripeness and sugar levels need to be carefully balanced by adequate acidity. The vine needs to have provided the grapes with sufficient moisture and trace elements to produce fine flavors. With such superb grapes, the skilled wine maker can then make a fine wine. As geographer Harm de Blij (1983:xi) noted:

> A great bottle of wine is a noble creation, a work of art as well as science, a triumph of talent and initiative, a progeny of natural environment and cultural tradition. As complex as a Monet landscape and as intricate as a Bach partita, such a wine is to the senses of smell and taste what painting is to the eye and music to the ear.

Fortunately, at least for all of us who enjoy wines, *Vitis vinifera* can adapt to a reasonably wide range of natural environments; however, each individual cultivar reaches perfection under a much narrower range of optimum physical conditions. In every important wine-growing region in the world, there are local variations in climate, differences in soils, and infinite patterns of exposure to sun, wind, and environmental hazards, especially frost—and all these factors must be reckoned with by winegrowers. As modern wine-making techniques have become the norm, more and more winegrowers are turning their eyes to the vineyards, seeking improvements there to match those already made in enology.

One notable aspect of viticultural evolution in America has been the gradual matching of grape varieties with the local environmental conditions in which they are at their best, with the optimum environment for a particular cultivar. This matching of grape varieties and microenvironments has been especially important for grapes that have proven to have special characteristics that result in distinctive, highly flavored wines.

Chardonnay, Cabernet Sauvignon, Pinot Noir, Riesling, and a handful of other cultivars top the list, but almost any of these grape varieties will show some preference for one place rather than another. Some, however, are more adaptable than others; whereas Pinot Noir, for example, can be extremely fussy about environments, Barbera and Chenin Blanc are less so. Resulting geographic distributions of cultivars reflect both environmental demands and market prices. For example, if two grapes find a particular environment optimum, then one may be grown more than, or even to the exclusion of, the other if the market sets high prices on one and low ones on the other. Locational patterns are further complicated, of course, by yields and production costs.

It is often desirable, especially in the warmer growing areas, to find grapes that will ripen without their acidity levels falling so low that only flat, unbalanced wines can be made from them. In California's Great Central Valley, for example, grapes such as Barbera, French Colombard, and Chenin Blanc have become important because they can adapt relatively well to the intensely hot summers (pay a visit to Fresno or Bakersfield in July for a taste of what this means) and still produce sound table wines. At the other extreme, winter hardiness is a necessity in many eastern viticultural areas, requiring growing American or French-American hybrids in place of *Vitis vinifera*; most of the latter are subject to damage or even death if winter temperatures drop below 0 degrees Fahrenheit for even a relatively short period of time.

The matching of grape varieties with their optimum local environments in the United States still goes on; it is likely to remain an ongoing process for the foreseeable future. What makes each portion of ground unique, at least with respect to viticulture, is its combination of weather and climate, soil, and exposure to natural hazards, a combination the French refer to collectively as *terroir*. Hugh Johnson (1994:22) has informed us that

> English has no precise translation for the French word *terroir*. Terrain comes nearest, but has a less specific, let alone emotive, connotation. Perhaps this is why many Anglo-Saxons mistrust it as a Gallic fancy. . . . It embraces the dirt itself, the subsoil beneath it, its physical properties and how they relate to the local climate—for example how quickly it drains, whether it reflects sunlight or absorbs its heat. It embraces the lie of the land: its degree of slope, its orientation to the sun, and the tricks of microclimate that spring from its location and its surroundings.

Weather and Climate

Hugh Johnson (1994:28) has argued (and most viticultural researchers would probably agree) that "the weather is the great variable in wine-growing. Every other major influence . . . is more or less constant, and known in advance." Even more directly, winegrower Philip Wagner (1974:106)—long associated with Boordy Vineyards, Maryland, and a great proponent of French-American hybrids—has stated with respect to viticulture that "climate has the last word."

Weather and climate are not synonymous, however. Climate is a long-term summary of weather, or the day-to-day conditions of the atmosphere. Thus, climate provides us with a general picture of where grapes can be grown, whereas weather variations from year to year affect each vintage (harvest year) in a different way. California weather varies less from one year to the next than either the weather of the eastern states or such European wine regions as Bordeaux, Burgundy, or the Rhine Valley in Germany; nonetheless, there are differences in the Golden State from one vintage to the next—every year is not ideal (despite a string of excellent vintages in the current decade). About vintages, wine writer James Laube 1995, p. 49) offers the following instructive comment:

> Is there such a thing as a perfect vintage? In theory, yes. It would go something like this: normal rainfall from late fall through winter, no frost and light (if any) rain in spring, an average to slightly below average-sized grape crop, mild, dry weather in the spring through summer months followed by a steady warming pattern in September—allowing a well balanced grape crop to ripen fully without the hindrance of rain.

Although California winegrowers may come closer to these conditions than their eastern American, French, and German counterparts, the ideal is often subject to considerable modification in any given year.

Macroclimates and Viticulture

At the world, or macro, scale, climate patterns are well established and change only slowly over long periods of time. Since 1900, considerable effort has been expended in order to classify climates and to delineate their regional geographical patterns. The most important climatic elements are temperature, precipitation, and seasonality. Wladimir Köppen developed a climate classification system that was based on the distribution of natural vegetation on the earth's surface. His system is still widely employed by geographers and included in many atlases.

The geographic range of *Vitis vinifera* (limited to temperate regions in both hemispheres) falls approximately between 30 and 50 degrees north latitude and between 30 and 40 degrees south latitude. Geographically, American and French-American hybrids fall generally within the same zone. Within these broad latitudinal belts, the climates that are most important for viticulture are those known as C climates, which are humid temperate or humid mesothermal in character. Some drier B climates are also used for viticulture, though irrigation is virtually always necessary; in the eastern United States, viticulture is also carried on in select locations within the microthermal D climates, often in sites where local conditions are tempered by the presence of water bodies that modify temperatures sufficiently.

In California, the Mediterranean climate (Cs) is of primary importance for viticulture, though in the Great Central Valley the drier B climates are also consequential. The Mediterranean climate is characterized by mild annual temperatures and dry summers. It can be subdivided into Csa, an interior Mediterranean phase with hot dry summers and cooler winter temperatures, and Csb, a coastal Mediterranean phase with cooler dry summers, more frequent summer fogs, and milder winter temperatures.

Most viticulture in the Pacific Northwest takes place in either of two very different climates. In Oregon's Willamette Valley, the climate, though broadly categorized as Cfb (marine West Coast), is very much like a cool Mediterranean climate. Most of Washington's grapes, on the other hand, grow east of the Cascade Mountains, in drier B climates. Although winters in Yakima and Walla Walla can be quite cold (with even a dusting of snow now and then), summers are warm and dry, allowing grapes to ripen nicely in the afternoon shadows of towering volcanoes. Cool temperatures at night help maintain good acid levels in the grapes, creating a harmonious balance between acidity and sugar levels, a balance that is essential for the production of fine table wines.

Most people quickly conjure up images of endless layers of naked rock and acres of cacti when they think about the southwestern United States, but they sometimes forget about another crucial variable that affects climate—elevation. Most of Arizona, New Mexico, and Texas falls within the B climate regimes (desert and semiarid). However, viticulture within these drier lands is carried on mostly at higher elevations and almost always with the use of irrigation. It is not unusual to find vineyards growing at elevations of 4,000 feet or more in these desolate lands.

From eastern Texas to the Atlantic, south of about 40 degrees north latitude, the climate is classified as Cfa (humid mesothermal).

Southeastern viticulture, including vineyards in eastern Texas, Arkansas, Missouri, Florida, Georgia, the Carolinas, and Virginia, occurs within selected pockets of this humid climatic regime—where temperatures are relatively mild, winters seldom severe (though freezes can be expected occasionally), and summer rainfall common (unlike the Mediterranean climate). Summer rains eliminate the need for irrigation but increase the threat of mildew and exacerbate harvest problems.

North of about 40 degrees north latitude in the eastern United States, the microthermal D climates prevail. Cold winters, snow, potential damaging spring frosts, and summer rains create a climatic combination that is not generally conducive to viticulture. Most of the major viticultural landscapes within this broad climatic region are located in places where local climatic modifications occur. Because water bodies heat up and cool down more slowly than the land surrounding them, they alter the nature of local microclimates considerably. Nonetheless, however, most of the viticulture in this cold macroclimate is characterized by American and French-American hybrids, though here and there particularly favorable sites allow the cultivation of selected *Vitis vinifera* cultivars. Vines need to be protected from winter freezes in many locations, and harsh winters can damage or even kill grapevines; viticulture in such regions is often a risky enterprise.

The southeastern shore of Lake Michigan, for example, provides conditions that allow viticulture in adjacent portions of Michigan and Indiana. Similarly, the south shore of Lake Erie is favorable to viticulture in adjacent segments of Ohio, Pennsylvania, and New York, as well as on Pelee Island. Two other New York locations are especially prominent in American viticulture (both capable of producing good Riesling and Chardonnay, and in some cases even Cabernet Sauvignon): the Finger Lakes region and northeastern Long Island (with well over 1,000 acres of vines, some on fields that once grew potatoes). Whereas the presence of several lakes in the former modifies local environments sufficiently to allow viticulture, the Atlantic Ocean, with its relatively warm Gulf Stream offshore, influences the latter. American *Vitis labrusca* and French-American hybrids are still predominant in the Finger Lakes region, but *vinifera* varieties are improving there; on Long Island, almost all of the vines are *vinifera*, led by Chardonnay.

Across the country, from Idaho to Massachusetts, there are a scattering of other areas in which viticulture is practiced. Almost always, grapes are grown within these states on very localized sites, typically where microclimates have been modified by rivers, lakes, or sheltered valleys.

Mesoclimates and Viticulture

When he was a professor in the enology department at the University of California at Davis, A. J. Winkler (1938) developed a classification of viticultural climates. His system employs a *heat summation index*, which is based on the sum of the number of degrees above 50 degrees Fahrenheit for each day during the growing season (generally from April 1 to October 31), to classify five different viticultural mesoclimates. Grapevines begin to grow when daytime temperatures reach 50 degrees Fahrenheit, hence the significance of that particular temperature in the calculation of the index.

If you have the necessary data, then the heat summation index is easy to calculate for a particular location. What you need is the average daily temperature for each day during the growing season. Then, for each day, you subtract 50 from that day's average temperature. Suppose that for August 1, for example, the weather station recorded an average temperature of 92 degrees Fahrenheit; you subtract 50 and end up with 42, the number of degree days for August 1. Now, do that for each day of the growing season (April through October), sum them all up, and you have, for that weather station, the value of the degree-day index devised by Winkler. Some examples of the heat summation index for locations in the United States include the following: Salem, Oregon, 2,030; Oakville, California, 2,300; Geneva, New York, 2,400; Yakima, Washington, 2,600; Santa Rosa, California, 2,610; Livermore, California, 3,260; Fresno, California, 4,680; and Bakersfield, California, 5,080.

Winkler's classification system (still widely used today, despite some notable shortcomings) includes five categories, each of which can in turn be related to the growing needs of specific grape varieties. These "mesoclimate regions" are more detailed in their distribution than the broader macroclimates, so they can be clearly related to both existing and likely future patterns of viticulture. We can now look at definitions and descriptions for each of these mesoclimates.

Region I

This climate has fewer than 2,500 degree days, as calculated by the heat summation index. *Vitis vinifera* vines require a minimum of approximately 1,700 degree days in order to produce mature grapes—most, of course, need more than that minimum. This is the coolest of Winkler's viticultural climates—akin to the growing climates in France's Burgundy (Beaune has an index of 2,300) and Champagne regions and Germany's Rhine and Moselle Valleys (where Trier has an index of 1,730, near the biological minimum for ripening grapes, mainly Riesling in this particular location).

Premium wine grapes are grown in this cool climate, with an emphasis on white grapes. Chardonnay, Riesling, and Gewürztraminer are white grapes that do especially well; Pinot Noir, a red grape, does well here also as long as it is not planted in the coolest parts. Hillside slopes and limited valley areas are used for viticulture in these cool growing regions. Where severe winters occur, growers may be limited to American or French-American hybrid cultivars. Occasional brutal freezes damage or even kill grapevines, as has occurred in places such as France's Chablis region and parts of New York state.

Region II

This viticultural mesoclimate (typical of Bordeaux in France, where the city of Bordeaux has an index of 2,519) is characterized by a heat summation index of between 2,501 and 3,000; it is extremely important for the production of grapes for premium table wines, both white and red. Although seemingly still on the cool side for viticulture, this growing climate provides excellent conditions for a wider variety of grapes than Region I, and grapes grown here include reds such as Cabernet Sauvignon, Merlot, Cabernet Franc, and Zinfandel. Among white grapes grown in this region are Chardonnay (grown in Region I as well, which is probably closer to optimum for that cultivar), Sauvignon Blanc, and Chenin Blanc. Valley floors are planted more extensively in these areas, though hillside and terrace vineyards are also common. Much of the middle Napa Valley is in this Region II growing regime, where Cabernet Sauvignon ripens to perfection. Between St. Helena and Calistoga, California, however, the mesoclimate warms into the Region III category. Parts of the Yakima Valley in

Washington state are examples of Region II as well, ripening Merlot to perfection.

Region III

Here we move into a somewhat warmer climate, with between 3,001 and 3,500 on the heat summation index. Warmer conditions favor grapes with higher sugar content; good-to-excellent table wines (including Cabernet Sauvignon and Zinfandel among red cultivars and Chenin Blanc among white ones) and excellent natural sweet wines can be produced here. In warm years, it may be a problem to make well-balanced table wines because of low acidity, though cooler years may be quite good.

Region IV

With between 3,501 and 4,000 degree days, this mesoclimate is generally too warm for the production of premium table wines, though it is often excellent for both natural and fortified sweet wines. In warm years, low acidity becomes a problem for table wines; sunburn is another frequent problem. However, careful trellising of the vines can help prevent the latter.

Among white cultivars, high-acid varieties such as Chenin Blanc and French Colombard are grown in this climate; they are used mainly in the production of inexpensive table wines. Somewhat unexpectedly and more in response to consumer preferences (and higher prices) than optimal environmental conditions, Chardonnay is appearing more and more in the area around Lodi and around nearby Clarksburg as well, both in California's Great Central Valley.

Red grapes, too, are planted in Region IV zones. Zinfandel grows in this mesoclimate, especially in the area around Lodi; today, much of it finds its way into White Zinfandel (for which grapes are often harvested at around 20 degrees Brix [pronounced "bricks," a measure of sugar content in a solution], so that acid levels can still be quite good). Barbera, Carignane, and Ruby Cabernet are other red grapes grown in this climate. These red grapes are often used in inexpensive table wines and in fortified dessert wines (which have become much less appealing to consumers in recent years). You have to wonder about the quality of some table wines, however, as you see more and more Chardonnay, Cabernet Sauvignon, and Merlot appearing around Lodi!

Most of the Region IV mesoclimatic regions are found either in the northern Great Central Valley or in southern California, from southern Los Angeles County on southward to San Diego County. At one time, the region that stretches westward from near downtown Los Angeles along the south side of the San Gabriel Mountains to Fontana was California's leading commercial viticultural area. Those days are gone, however; suburban housing tracts, shopping centers, and smog have driven most vineyards out of the region.

Region V

The Region V mesoclimatic region is extremely large and includes much of California's Great Central Valley between Redding and Bakersfield as well as portions of southern California. Hot summers in this mesoclimate produce grapes with high sugar content, but low acidity is a constant problem. For that reason most grape varieties here need to be high-acid varieties. Modern wine-making techniques allow wine makers to produce clean, drinkable table wines if the right grape varieties are chosen. Dessert wines such as ports and sherries can be excellent. The region also grows table and raisin grapes in considerable quantities. One of those, the Thompson Seedless grape, finds its way into many inexpensive white wines as well—in the mid-1990s around 8.0 percent of California's grape crush for wines came from that single cultivar.

Summary of Winkler's Classification

As a general "rule of thumb," we can say that the best premium table wines come from high-quality grapes grown in the cool Region I and II climates, whereas the best dessert and fortified wines are produced from grapes grown in Regions IV and V. Region III is transitional; it is especially good for natural sweet wines, though it often produces very good table wines as well, especially in cooler years.

It is useful to remember that the warmer a growing climate is, the more sugar the grapes grown there will have and the lower their acid levels will tend to be; the best table wines need a careful balance of these two elements. In California, cooler years tend to produce the best table wines, but in the more northerly premium European and American wine regions the opposite is true. Whereas the eastern American and European wine makers worry about having sufficient sugar levels in their grapes, California wine makers worry more about having sufficient acid levels.

Microclimates and Viticulture

Microclimates are highly localized, so many different ones may be found within any one mesoclimatic region. Just as you find different growing environments in different parts of your own yard (full sun, part sun, shade), each of which may support its own array of plants, so we can also easily imagine that within any of Winkler's mesoclimatic regions there are local variations in weather and climate. Thus, individual vineyards within a specific Winkler mesoclimatic region may experience differences in the amount and intensity of sunlight received (because of the direction a particular slope faces on a hillside, for example), variations in rainfall and runoff (again often a consequence of local topographic variations), and dissimilar exposure to fog or frost. Valley bottoms, for example, tend to be more frost prone than hill slopes because cold air drains down slopes into low-lying areas. Also, hill slopes may stand above local shallow fogs.

Winkler's mesoclimatic classification outlines the broad geographic patterns of viticulture in the United States, but we should remember that considerable local variations do exist. Good viticulturists and wine makers (and there are many in the nation!) are aware of these differences and sometimes exploit them to advantage; knowledgeable wine consumers are becoming more aware of them also. At the same time, however, many of the smaller microclimatic variations are averaged out when grapes from larger vineyards are crushed together, as is often the case. Continued experimenting will lead to ever-improved matches between cultivars and their optimum environments, giving wine lovers much to look forward to as American winegrowing matures.

Soil

Vitis vinifera is a deep-rooted plant with roots that can extend to depths of twenty feet or more. In that dank and dark subterranean realm, the roots seek water and nutrients so that both water and a variety of minerals can be carried up to the vines and into the grapes. American viticulturists have tended to assign a lesser role to soil than have the French in explaining grape and wine quality. According to Winkler and his colleagues in their classic *General Viticulture* (1974:71):

> Grapes are adapted to a wide range of soil types. True, one finds a decided preference for certain soil types in nearly every grape-growing district. Nevertheless, when all soils used for growing the various kinds of grapes in the many different grape-producing regions of the world are compared, one finds that they range from gravelly sands to heavy clays, from shallow to very deep, and from low to high fertility. One should avoid heavy clays, very shallow soils, poorly drained soils, and those that contain high concentrations of salts of the alkali metals, boron, or other toxic substances.

Similarly, geographer Harm de Blij (1981:77) noted that "to predict how a vine will grow in a particular soil is even more difficult than to forecast its successes under certain climatic conditions." Despite these statements, many wine makers and consumers continue to debate the effect of *"terroir,"* arguing that grape quality is notably affected by soils—and scientific evidence provides them with no help. Noted wine writers Hugh Johnson and James Halliday (1992:20) have gone so far as to suggest that "all scientific attempts to show a specific translocation of minerals or other substances from the soil, which then impact directly on the flavour of the grape (or the wine) have failed."

However much contention still exists about the precise contribution that soils make to shaping the final product, most viticulturists do agree that the success of viticulture is not directly dependent on the inherent fertility of the soil. According to James Halliday (1993:18), "If the concept of terroir is properly understood, and if one ignores the unscientific propaganda trotted out by public relations personnel on both sides of the Atlantic, there is a surprising measure of agreement: soil structure (and to a lesser degree soil texture) is the most important factor; soil type (in terms of mineralogical and organic composition) is the least important."

Vines grown on hillsides with poor soils have been known to produce splendid wines—and numerous wine makers even argue that vines must be "stressed" in order to produce their best fruit. Many readers will recognize the *Graves* appellation that appears on some French Bordeaux wines; translated into English, the word means "gravel," which is hardly of great inherent fertility but offers ease in the downward movement of groundwater. Fertile soils are best suited for raisins; less fertile soils produce the best-balanced premium wine grapes. For wine drinkers, this has been fortuitous because food crops can be grown on the best soils and grapes moved onto the more marginal ones.

Deep soils with good drainage (perhaps the most important soil quality for viticulture) are ideal for most premium wine grapes—soil structure and texture, not inherent fertility, are the keys to viticultural success. Well-drained soils allow the roots to go deep, where they can tap groundwater (even when surface conditions appear hopelessly dry, as they do in much of California, eastern Washington, Arizona, New Mexico, and western Texas during the summer) and a wide variety of minerals. Deep roots also protect the vines from temporary droughts; they are essential in unirrigated vineyards, where summer rains may be virtually nonexistent and are not really wanted.

Drainage conditions are related to soil texture, which is a measure of the particle-size composition of the soils. Texture is normally expressed in the percentage of particles present in three sizes: sand, silt, and clay. Texture affects not only drainage but also the quantity of water that a soil can hold and the ability of the soil to allow circulation of both water and nutrients through the soil.

Soil texture (and to some extent color) is also related to the heating and heat-retention characteristics of soils, thus affecting microclimates. Hugh Johnson (1994:83), writing about the relationship between soils and wine quality in Bordeaux, noted that "heavy clay or sand which drains badly are the least propitious components for wine: gravel and larger stones are best." He went on to say (1994:83), "Add to this the way stones store heat on the surface, and prevent rapid evaporation of moisture from under them, and it is easy to see that they are the best guarantee of stable conditions of temperature and humidity that a vine can have."

Premium wine grapes, then, are likely to be at their best in soils that are deep and coarse in texture. In many of America's major viticultural regions, the best soils are alluvial valley soils, which are found at intermediate elevations between the fine-textured and often poorly drained valley basin soils and the higher soils of terraces and steeper mountain slopes. Excellent vineyards are also found on terrace and upland soils, though there are many problems and increased costs when grapes are planted on steeper slopes—but high quality, of course, can help offset the higher costs. As winegrower Justin Meyer (1989:84) has commented, "Every time he plants a new vineyard, a vintner or grower must match a grape variety to the climatic region and the soil type in a fashion that will produce superior fruit and wine."

Environmental Hazards

Given proper growing climates and decent soils, it is possible to successfully plant fine vineyards and produce fine wines. However, viticulture can also be affected by a variety of environmental hazards. In the United States, frost, hail, wind, unseasonal rain, and fire can all produce damage to grapevines, as can birds, rabbits, rodents, deer, and the many pests and diseases that strike specifically at *Vitis vinifera*. Here we look at some of these in more detail, but all are variable in their presence and severity and are generally limited in their geographical extent.

Frost

Although killing frosts are not common in California's viticultural regions, frost can occasionally be a major problem, especially if it strikes during a vine's budding or flowering stage. Vineyards in the northeastern United States, of course, have far more threats from spring frosts because the winter climate is so much colder there; thus, winter hardiness becomes important in the selection of vines. Damage during early stages of growth may lower or even destroy the year's grape yield.

Topography is important in the geographical distribution of frost damage. At night, as the ground cools, cold air drains down from the hillsides and collects in the valley bottoms; low-lying vineyards are therefore especially likely to experience frost problems. Wind machines and overhead sprinklers can provide at least some frost protection.

Hail and Rain

Because of the dry summer Mediterranean climate in California, hail is seldom a problem in most vineyards there during the growing season (however, it has occurred on occasion, including the spring of 1995); in midwestern and eastern vineyards, it can wreak havoc more regularly. When it does hail, damage is usually sporadic and highly localized; seldom does it affect a wide area. Because storms are shunted far northward in western North America during the summer months by the presence of high pressure and the northward shift of the polar jet stream, high winds and unseasonal rains are seldom problems, though the unusual El Niño conditions in 1982 and 1995 created rain problems in several areas. Many Napa vineyards were flooded in 1995, for example, and the continued rains during the spring of that year resulted in a reduced harvest in the fall (helping to drive up wine prices significantly).

Rain is especially damaging to thin-skinned grape varieties such as Zinfandel. Too much moisture encourages mildew and rot, and rain at harvest time can be especially difficult to deal with. Summer rains are much more likely in the eastern United States, where tropical moisture from the Azores High moves toward the continental interior, resulting in much convective rainfall and high humidity.

Fire

In dry summer climates, brushfires can be severely damaging. The Lexington fire near Los Gatos, California, for example, burned for three days in summer 1985, though its actual damage to vineyards was minimal. One result was the naming of Wildfire Vineyards, a twenty-five-acre vineyard that, before the fire, was unnamed. In 1996, a summer fire near Glen Ellen, California, badly damaged more than forty acres of grapevines and threatened at least one winery. Although care can be taken within the vineyards, most fires start on adjacent lands.

Vineyard Pests

Vitis vinifera is subject to attack by a variety of pests and diseases, and a few of the most important are discussed in this section. Vineyard pests in viticultural regions run the gamut from hungry deer and birds to leaf mites, leafhoppers, and phylloxera, the tiny yellow root louse that has destroyed vineyards around the world since the 1860s. Healthy vineyards are obviously one of the prerequisites for producing quality wines.

Phylloxera

Probably no other disease or pest has wreaked such havoc in the world's vineyards as this tiny bug. Until the early 1980s, phylloxera was not thought to be a significant threat to most American vineyards because earlier lessons had been well learned. Wherever phylloxera was thought to be a potential problem, vines were planted on resistant American rootstocks. Unfortunately, the widely used AxR#1 rootstock (a hybrid between *Vitis rupestris* and a *Vitis vinifera* cultivar, Aramon), which had proved to be resistant to phylloxera, came under attack in some California vineyards by a mutant form of this tiny bug, which was subsequently identified by researchers at the University of California at Davis, as phylloxera Type B (the original has now been labeled Type A). During the 1990s, the spreading infestation of phylloxera has made it necessary for winegrowers to make widespread replantings of vineyards in Napa, Sonoma, and other areas in California.

This has taken a considerable toll on the profit margins of numerous wineries, though it has at least occurred at a time when the demand for California wines is high. Replanting is expensive—overall costs in Napa and Sonoma Counties alone may reach $1 billion (Sullivan 1996). Furthermore, the search for new resistant rootstocks continues unabated; several are currently being used, with Teleki 5C perhaps the favorite for now. Interestingly enough, one benefit of the replanting has been that varietal selection is being done more carefully; new trellising systems are also being introduced, and vines are being planted at considerably higher densities than they had been earlier. In Napa, for example, earlier vineyards typically contained about 450 vines per acre; today, some are being planted with more than 2,000 vines per acre. Yields are likely to go up; and some wine makers believe that quality will improve as well, though that seems contrary to what we often hear and read about the relationship between low yields and high quality. The proof, of course, will be in the bottles of the future.

Pierce's Disease

Another serious problem for grapevines is Pierce's disease, for which there is no known cure. It is endemic in parts of the southeastern United States from Florida to Texas. Its cause is a bacteria, *Xylella fas-*

tidiosa, that thrives in reeds and other riparian vegetation and is spread into vineyards by tiny flying insects known as "sharpshooters." Although it is not now widespread in American vineyards outside of the southeastern region, it was responsible for completely destroying the vines that were planted by Germans in Anaheim, California, during the nineteenth century. In the mid-1990s, Pierce's disease is becoming problematic in parts of Napa and Sonoma Counties in California.

Unwanted Grape and Vine Lovers

Ripe grapes appeal not only to humans but also to quite a few other members of the animal kingdom. Birds are especially troublesome, swooping down on vineyards and eating more than their share in record time. Tender shoots and young vines are tasty as well; thus, rabbits and other rodents are especially harmful to young vines, which are often surrounded by milk or juice cartons, called "bunny boxes," for protection. Browsing deer also take a toll, though they are much harder to keep out, especially since fences are not common in most vineyard areas. High fences around vineyards in California's Sierra Nevada foothills are becoming more common, however, as grape growers attempt to keep these hungry denizens away from ripening crops.

Managing Environmental Problems

It is possible, within reason and budgetary constraints, for viticulturists to mitigate at least some of the natural problems that concern them. Without going into detail, a brief survey provides an introduction to some common practices.

Irrigation protects vineyards from inadequate natural water supplies and helps regulate water delivery. Its need is most obvious in the driest vineyard areas, but irrigation is common in young vineyards even in California's Napa and Sonoma Counties. Young vines are more susceptible to seasonal droughts than mature vines because their roots have not yet grown deep enough to tap into underground water supplies. The most common irrigation techniques are to use overhead sprinklers, which generally stand above the vines, and drip irrigation systems, where tiny hoses near the ground carry water to each individual vine. The latter have become increasingly more widespread during the last decade or so because they are so efficient at delivering water directly to the vines, thus reducing irrigation costs.

Frost damage can often be mitigated also, either by using overhead sprinklers or wind machines, though the latter, most of which are large propellers scattered throughout vineyards, are less efficient. Fire damage can be minimized by providing fire breaks. Deer fences can discourage deer, and "bunny boxes" protect young vines from becoming meals for rabbits and rodents. Nets (though expensive) and field cannons that fire noisily (but harmlessly) at repeated intervals can help minimize bird problems.

Pesticides and herbicides are used in the vineyards to fight everything from weeds, which use scarce moisture, to mildew and mites. Phylloxera is controlled by planting the phylloxera-resistant rootstocks of non-*vinifera* vines from the eastern United States and then grafting *Vitis vinifera* onto that rootstock. This is common practice throughout most of the world's major wine-growing regions. However, there are some American vineyards that have been planted directly with *Vitis vinifera* on its own roots. How these vines fare in the long run remains to be seen, but already phylloxera has become a serious problem, especially in Napa, Sonoma, and other parts of California, as noted already.

Other pests are increasingly being controlled by nonchemical means when possible; there is an ongoing "greening" of viticulture, a movement toward the development and use of practices that will be sustainable in the long run. Sustainable agriculture (including viticulture) attempts to combine three basic goals: environmental health, economic profitability, and social and economic equity. Its underlying principle is the meeting of current needs without compromising the ability for future generations to provide for their needs. Making the transition to sustainable agriculture is going to be a gradual, and perhaps sometimes painful, process; however, viticulturists are among the vanguard in this transition. Ultimately, everyone from farmers to final consumers is going to have to play a role if sustainability is to be attained.

Perches for hawks and owls are being added to vineyards to encourage these rodent-eating birds to stick around. Cats are becoming more common residents of wineries and winery buildings as well—their love of mice and other rodents recommends them highly. As writer Jo Diaz (1995:34) commented, "the winery cat is to the winery and vineyards what the sheep dog is to the flock and meadows." In addition to the use of animals for pest control, growing different types of ground cover below and between vines encourages helpful insects to stay around and make meals of harmful ones, cutting both the costs and environmental threats of using chemicals. Integrated pest management (IPM) has become much more common among grape growers in recent years, representing one part of the shift to sustainable agriculture that seems so important as we approach the new millennium.

Summary

Overall, the United States offers a number of amenable environments for wine growing. High-quality premium wines can be produced in the cooler regions (sometimes with American or French-American hybrids), and good drinkable wines (perhaps the best in the world in this category) can be produced in large quantities in the warmer regions, especially when high-acid grape varieties are selected for planting. California's mild winters minimize freezing problems, whereas eastern growers are less fortunate; early frosts, though possible, are not as severe a problem as they are in places like Germany's Rheingau or France's Burgundy or Champagne.

Plenty of sun during the summer (especially in western vineyards) means that grapes here usually ripen to maturity without difficulty, though at the same time maintaining acid levels becomes a problem in the warmer viticultural climates such as those of California's Great Central Valley, especially in warm years. It is little wonder, then, that California continues to produce more than 90 percent of the nation's wines, the best of which can compete with any in the world. Washington state is becoming a formidable competitor, however, and more selective successes can be found from Oregon's Pinot Noirs to the excellent sparkling wines, Rieslings, and Chardonnays of Ohio, Pennsylvania, and New York.

4

$\mathcal{A}$merican Wine Making Comes of Age

1992
CABERNET SAUVIGNON
NAPA VALLEY

$\mathcal{E}$SSENTIALLY, WINE IS A SIMPLE BEVERAGE; it can literally make itself. It is consumed by millions every day with little more thought given to it than to the bread that typically accompanies it. This wholesome beverage is far more often consumed from water glasses than fine crystal, with no apparent loss in satisfaction for the average wine drinker, despite what many wine writers and glass manufacturers would have you believe. Today, wine makers are able to make acceptable wines with relative ease, though the best wines require far more than that statement suggests. Modern American wine making ranks among the best in the world; American enological innovations have diffused widely throughout the wine-growing regions of Europe and are making their way into the Southern Hemisphere as well, from South Africa and Australia to Chile and Argentina. The quality of wines from all of these regions has increased significantly during the last decade or so.

Within the United States, new technologies have been embraced by most wineries during the past thirty years, though capital scarcity has slowed adoption of some new equipment by many small wineries. New technology has created a wine-making atmosphere that is much more science than art, though wine makers still make final decisions that may result as much from intuition as from analysis, from careful tasting and thinking rather than from computer readouts of test tube analyses.

Wine must assuredly have seemed mysterious to those first early "wine makers." They had no scientific explanation for what they observed: Grapes, left to their own devices (the natural combination of grape sugar and wild yeast initiates fermentation without any help from humans) in a crude container, could produce a special and delightful beverage—wine—and this must have helped make life better for many at an early date. We do know, though, that the engaging mysteries of those ancient products of the wine maker's art became indelibly intertwined with religious experiences at a very early time. Bacchus and Dionysus are still invoked today, reminders of the esteem with which wine was held by many ancient peoples.

In our current age of modern science and technology, much of the mystery that once cloaked this unique beverage has been removed. Within the last century, almost every operation in wine making has become clearly understood and, when possible, has been refined and modified to produce the best possible product. This is not to say, however, that science alone has replaced the wine maker's art; subjective decisions still have to be made at every step of the process, from decisions about planting and tending vineyards through deciding when to harvest and how long to ferment to deciding which final blends reach the bottling lines.

Grapes Become Wine

Scarcely more than one century ago, French scientist Louis Pasteur (linked forever to milk, via "pasteurization") discovered that fermentation could be simply defined as life without air, that it was an "anaerobic" process, as opposed to "aerobic" processes (including that popular exercise). In the fermentation process, sugar is converted into alcohol and carbon dioxide by the action of yeast; the alcohol is retained and the carbon dioxide harmlessly escapes (except when sparkling wines are being made). It is from fermentation that the living yeast receives its energy. Without this energy, the yeast could not reproduce; without yeast there would be no wine (and no beer or leavened bread, for that matter). And some people, Hugh Johnson and Robert Mondavi among them, might go so far as to say that without wine there would be no civilization as we know it.

In ancient times, wine makers depended entirely on the wild yeasts present on the skins of the grapes to initiate fermentation. Today, however, the types and amounts of yeasts are usually controlled with considerable care, though occasionally wine makers still encourage wild yeasts to do the job. As enology professors Maynard Amerine and Vernon Singleton (1977:74) noted, "The biochemistry of fermentation not only beautifully illuminates and clarifies the ancient art of making wine, but also explains and makes possible a calculated control of many of the 'mysteries' which baffled the artisan winemaker." They added, "Aesthetics, however, is still a part of (and should not be displaced from) wine production and wine appreciation, but mystique should not remain if knowledge can be substituted." Common strains of wine yeasts used in American wine making today include Montrachet, Pasteur Champagne, California Champagne, and Epernay 2; the choice of yeast depends both on preferences of individual wine makers and on the cultivars to which they are adding it.

From the planting of the vines to the bottling of the final product, a majority of American winegrowers apply the most modern knowledge and sophisticated equipment to make the best wines that they possibly can. The best contemporary wineries have well-equipped laboratories, or they send wine samples out to laboratories for analysis if they do not. Wine makers, especially at the larger and better-capitalized wineries, are as likely to look and talk like chemists as they are to resemble the romantic characters that people often envision them to be, the "little old wine makers" of marketing fame, such as in the old advertisement for Italian Swiss Colony. Outside their labs, however, many have achieved "star" status in the 1990s. Wine makers Paul Draper, Tim Mondavi, Randy Dunn, Heidi Barrett, Warren Winiarski, Randall Graham, and Helen Turley, for example, are as well known to lovers of wine and food as are chefs Julia Child, Wolfgang Puck, Marcella Hazan, and Alice Waters.

Different types of wines require different production techniques. We will consider only the fundamental steps here in the production of table wine (white, red, rosé, and blush), sparkling wine (champagne), and fortified wine (port and sherry). In each case *only* the basics are discussed below. (Interested readers in search of more detail and depth are encouraged to consult references such as Jeff Cox [1985] or Emile Peynaud [1984].)

White Table Wine

White table wine is produced by fermenting only the juice of the grape (which is white—or clear—even in most red grapes), without much extraction of solids and without the skins of the grapes. Although it sounds simple enough, making consistently good white wines is not always easy. Writing about the production of white wines, Amerine and Singleton (1977:121) have said that "when well made, they have a delicate fruity yet vinous odor (with the appropriate varietal notes) and a light straw to bright medium-gold color, with no muddiness or brown and no off-flavors."

Care must be taken to harvest grapes when they are at exactly the right stage of maturity, generally with sugar levels around 20 to 23 degrees Brix. Brix is a scale for measuring the sugar content of a solution; it assesses the weight of sugar in the juice, or "must," as a proportion of the total weight of the solution. The most common tools for measuring the sugar content of a must are hydrometers and refractometers (for details about their use see Amerine and Ough [1980]). In simple terms, if we have 100 pounds of must, and it contains 23 pounds of sugar, then the must would have a sugar content of 23 degrees Brix (most, but not all, of the sugars—mainly glucose and fructose—in the must are fermentable).

A careful balance between sugar and acidity needs to be maintained in order to produce good table wines. Overly ripe grapes not only lead to an imbalance between sugar (too much) and acid (too little) but also to alcohol levels that may also be too high for a table wine. In addition, low acidity results in "flatness" or "flabbiness," leading to wines that are insipid. Overripe grapes may also give wines an undesirable "raisiny" quality, at one time common in many California Zinfandels, especially those made from grapes grown in the Great Central Valley.

If the must is fermented dry (that is, until there is no fermentable sugar left), then the alcohol content of the finished wine can be calculated by multiplying the Brix measurement of the unfermented must by a factor of 0.55. For example, if a must measured 23 degrees Brix and was fermented dry (normally defined as having less than 0.1 percent residual sugar in the finished wine, though sugar probably remains undetectable for most people up to about 0.5 percent), it would then have an alcohol content of around 12.6 percent, a typical figure for a dry table wine.

Once grapes for white wines have reached their desired maturity level, it is important that they be harvested promptly and moved as expeditiously as possible from the vineyard to the winery. There, they go immediately to a crusher-stemmer, which both breaks the skins and separates the grapes from their stems. Allowing the grapes to sit around after they are harvested courts disaster; evaporation, bacterial activity, and oxidation are potential hazards.

From the crusher-stemmer, the crushed grapes are moved to the press (in modern wineries, bladder presses are commonly used today), where the juice is separated from the pulp, seeds, and skins. The juice, or "must" as it is generally known at this point, is treated with sulfur dioxide (which is why virtually all commercial wines contain sulfites, a fact now duly noted on wine labels), and is then allowed to sit for a few hours. Sulfur dioxide inhibits browning, slows oxidation, and helps to control the growth of undesirable organisms, especially wild yeasts and bacteria. Care must be taken, however, to mix the sulfur dioxide carefully throughout the must and to use it in amounts that are sufficient for the purpose but not so great as to become discernible in the finished product. Too much sulfur can be detrimental to a wine's quality, resulting in undesirable sulfurous aromas and flavors. Sulfur may be especially detectable when a bottle is first opened.

The must then enters fermentation vats, where a carefully selected wine yeast is added and fermentation begins. In America today, most white wines are fermented in temperature-controlled stainless steel fermentation tanks at relatively low temperatures—between 45 and 60 degrees Fahrenheit. At these lower temperatures, fermentation takes three weeks or more, but that time is well spent because white wines fermented at low temperatures preserve their natural fruitiness much better than those fermented at higher temperatures.

Despite the overwhelming popularity of temperature-controlled fermentation in stainless steel tanks for most white wines, some producers follow other procedures. For example, a number of premium wine producers like to ferment at least a portion of their Chardonnay in small oak barrels (such as French Nevers or Limousin) in order to increase its complexity. However, careful judgment must be exercised by the wine maker or the resultant wine may be unbalanced because of excess "oakiness," a common occurrence in many California Chardonnays since the 1970s. Too much oak can detract from the fruit's natural flavors and aromas, and heavily oaked wines may be hard to match with foods. Writing about Chardonnay, for example, wine writer Dan Berger (1995) said, "I prefer fruit flavors and aromas to those arising from production techniques. . . . I look for wines that are made simply, so the fruit has a chance to assert itself." Others, of course, want to taste oak in their Chardonnay; there is a sufficient variety of different styles out there to provide something for almost everyone.

After fermentation, white wines are typically racked (removed from the solids, or lees, that have settled to the bottom of the fermentation tanks), fined (to remove remaining sediments that are suspended in the new wine), and perhaps cold-stabilized (to prevent the formation of tartrate crystals in the bottle). The wine then goes into oak barrels, stainless steel tanks, or bottles, depending on the variety and style of the grape and wine and the discretion of the wine maker.

Fruity white wines such as Riesling and Chenin Blanc are usually ready for stainless steel and then bottles, though occasionally Chenin Blanc may receive a short time in oak barrels. Chardonnay, on the other hand, is generally destined first for some time in oak. Wine makers generally use barrels made from such French oaks as Nevers and Limousin. However, American oak is gaining in popularity, be-

These oak barrels are being "toasted" to be made ready for use. Photo courtesy of Seguin Moreau USA, Inc.

cause of its lower cost and improved cooperage techniques, most of which have been brought to the United States by French cooperage firms. After aging in oak, the Chardonnay will be ready for bottling.

Before bottling, however, some wine makers blend different lots of Chardonnay—say, some aged in Nevers and others in Limousin—for even more complexity of flavors and aromas. For the most part, those white wines that do not go into oak are ready to drink quite young and will be short-lived; they should be enjoyed immediately for their inherent freshness and forthright fruitiness. Those that gain some complexity from oak are going to mature somewhat more slowly. Many American Chardonnays will improve for at least two or three years in bottles, though most are not going to be long-lived.

Red Table Wine

The great American wine authority Leon Adams, who died in 1995 in his ninetieth year, used to say that all wines would be red if they could. Red wines received a significant boost in popularity when the television news program *60 Minutes* did a segment in 1991 (and a follow-up in 1995) touting the health benefits of red wine. Always eager for the "quick fix," Americans quickly began consuming more red wines.

Grapes for red table wines are generally harvested at 21 to 23 degrees Brix. Occasionally, Zinfandels have been made from riper grapes, sometimes creating alcoholic "monsters," though such wines are currently out of favor with most consumers. Unlike the process for making white wines, red grapes do not have the juice immediately separated from the skins after they are crushed. Tannins and coloring pigments for red wines come mainly from the skins; the juice of most red grapes is actually clear. As with white wines, however, the grapes are treated with sulfur dioxide during or immediately after crushing.

Once in the fermentation tanks, the must is inoculated with yeast, and fermentation then begins. Unlike white wines, red wines are still not normally fermented at low temperatures, though it is important to see that the temperature of the fermenting must does not rise above about 80–85 degrees Fahrenheit. Because of the higher fermentation temperatures, red wines complete their fermentation more rapidly than white wines, usually in about one week.

Before fermentation ceases, however, the wine maker must closely monitor the extraction of color and tannins. When desired color and tannin levels are reached, the must is moved to the press so that it can be separated from the skins and other solids. Fermentation may continue after pressing if it has not already finished.

The completed red wine is then racked for clarification and moved into containers for aging. These are generally small oak barrels of French, American, or even German or Yugoslavian origin, according to the tastes and financial resources of the wine maker. Sometimes more neutral (in the sense that they have little or no effect on the taste of the finished wine), redwood or stainless steel vats may be employed. There has been a general tendency to reduce the amount of time that American red wines are aged in oak barrels, partly in response to consumer preferences for lighter, less tannic red wines that can be consumed while they are still relatively young. Most American wine consumers drink their wines soon after they purchase them.

We should at least briefly mention the increasingly popular *nouveau* (literally, "new") wines. Long a tradition in France's Beaujolais region, these wines, though red, are released in the November immediately following the vintage. In other words, in the fall, grapes are harvested, vinified, and bottled for release in November of that same year—they are aged not at all. More and more French Beaujolais is made in this fashion because it is a real "cash cow" for the producers. More California producers are entering the fray as well, making products that certainly rival the French in quality and usually beat them in price. In recent years, even the Italians have entered the nouveau wine market.

Nouveau wines are produced by a process called carbonic maceration, which involves processing the grapes for several days under a pressurized carbon dioxide "blanket." The main effect of this processing is to soften the tannins in the must, thus allowing the red wine to be drinkable within a few weeks of the harvest. In France, the release of the first Beaujolais wines in November is cause for celebration; it provides the first indication of the quality of the vintage (and, of course, one more good reason for Parisians and others to revel in the pleasures of the grape). Nouveau wines are often pleasant, even "quaffable," but they are short-lived and should be consumed within a few months of their release. Their appeal is their youthful fruitiness, not their ageworthiness. In California, various red grapes have been used to make nouveau wines, including Gamay, Gamay Beaujolais, and Zinfandel.

Rosé Table Wine

Novices often believe that rosé wines are made by combining white and red wines in appropriate proportions. Although this has undoubtedly been done by some American wineries, most rosé wines are not made that way. Rather, rosé wines are made from red grapes. The key to the depth of color is the amount of time that the crushed grapes are kept in contact with their skins before separation in the press. Typically, that time ranges from twelve to twenty-four hours, time enough for fermentation to get under way and for a nice pinkish color and distinctive flavor to be extracted from pigments in the skins. Then, the must is separated from the skins with a wine press and vinified as if it were a white wine.

Rosé wines are virtually never aged in oak; their winsome charm comes from their delicate color and light, youthful fruitiness. The best rosés take considerable character from their parent grapes, especially from such distinctive cultivars as Cabernet Sauvignon, Gamay, and Zinfandel. Grenache has long been popular for rosé wines also, though it is often less distinctive than the varieties already mentioned.

Blush Table Wine

Two decades ago, blush wines were virtually unheard of in the United States. However, in recent years their popularity has been phenomenal. Technically, they are no more than very light rosés, what the French have called *vin gris*, taking only a "blush" of color from the grape skins during pressing.

Beginning in the mid-1970s, it became apparent to most wine makers that the "wine boom" in the United States was mainly a white wine boom, but grape plantings in the late 1960s and early 1970s had been mainly red cultivars. In order to capitalize on changing consumer preferences without vast replantings of white grapes, producers (led by wine makers such as Bob Trinchero at Sutter Home) turned to making blush wines.

As noted already, the juice of most red grapes is clear; red wines take their color (as well as much of their flavor and tannin) from the skins. Thus, a nearly white wine can be made if the juice from red grapes is separated from the skins as quickly as possible and vinified like a white wine. However, the juice almost always retains a light kiss of color, resulting in the term "blush." Such wines are delightfully appealing to the eye, with colors ranging from delicate pink to a very light salmon. Blush wines are generally off-dry; they typically contain 1.5 percent to 2.5 percent residual sugar. They are soft and drinkable, full of fresh fruit flavors, tasty by themselves or with light meals, and they have been successful beyond anyone's wildest expectations. They have also attracted legions of new wine drinkers, even if they lack even the slightest trace of snob appeal.

Although they can be made from many varieties of red grapes, blush wines made from Zinfandel grapes perhaps best illustrate just how enticing these wines can be. Zinfandel is the most widely planted red grape in California; currently there are nearly 40,000 acres of it, concentrated principally around Lodi. The berrylike flavors associated with this grape carry over nicely, though lightly, into the blush wine versions, giving them an aroma and flavor that is often reminiscent of fresh strawberries. Slow temperature-controlled fermentation in stainless steel tanks ensures that the fresh fruitiness of this grape is captured and kept for the consumer's drinking pleasure. These wines are made to be consumed young. Blush wines have also been made from Pinot Noir and Cabernet Sauvignon and occasionally from other red grapes as well, including Merlot, but it was White Zinfandel that made them what they are today.

Bubbles in the Wine

Confusion is common in the terminology that prevails in America's sparkling wine industry today. Consumers often differentiate between champagne and sparkling wine, but not in the way that many wine writers and wine makers have encouraged them to do; it seems a case of good intentions gone awry.

For many wine writers (especially if they have any French connection), the term "Champagne" (with a capital "C") is reserved only for the sparkling wines of Champagne, that carefully delimited and controlled region northeast of Paris that includes such well-known cities as Reims and Epernay. This region is still indisputably the home of great Champagnes, though American versions get better every year (at costs that are much more attractive right now than their French relatives, thanks to our feeble dollar). As the French would have it, *no* other sparkling wine is Champagne, nor should any ever be called by that sacred name. It was within France's Champagne region—which gives its name to the beverage, of course—that sparkling wines were discovered, improved, and ultimately perfected. It was there also that Dom Perignon, in the abbey at Hautvillers, upon first tasting Champagne, reputedly heralded, "Look! I'm drinking stars." Whatever the veracity of the legend, Dom and his colleagues went on to make numerous improvements in the production of Champagne, and his name has been immortalized by Moët and Chandon in their world-renowned Dom Perignon, one of a very few superpremium sparkling wines produced in the Champagne region.

Why the confusion about Champagne? It occurs mainly among the many consumers who have read or heard too little, a fault we should quickly forgive them for, because you really shouldn't have to read and learn about a beverage in order to enjoy it (though it may help you enjoy it even more if you do, of course). To a majority of these people—and their numbers are legion—Champagne is any sparkling wine, whether it is the least expensive American version or Dom Perignon. For this reason, the term is still used by most American sparkling wine producers, despite vociferous protests from the French.

To the great mass of champagne (a generic term, with a small "c") consumers (most of whom still associate the drink with celebrations and special occasions, from weddings to winning locker rooms— where bottles are more often sprayed about than consumed)—the term "sparkling wine" has little or no meaning. Worse yet, it often suggests something like apple or grape juice. Many consumers do not even realize that champagne is a wine, which is one of its identity problems. In their minds, then, champagne and sparkling wine must be different—the latter perhaps only an imitation of the former—but they are convinced that it is champagne that they want.

Sparkling wine refers to any wine in which carbon dioxide bubbles are trapped in sufficient quantity—and in that sense, all champagne is sparkling wine, but not all (or even most) sparkling wine is truly Champagne, if we are willing to reserve that term only for the prestigious French product from the Champagne region. However, many American producers continue to call their sparkling wines Champagne—and as some might say, so much for the French! Today, of course, sparkling wines are made in most of the world's wine regions. Countries in the European Union cannot use the term Champagne for any of their sparkling wines, so they must use their own names: Cava in Spain, Sekt in Germany, and Spumante in Italy are examples.

In America, there are now some sixty or more producers of sparkling wines, including several that are owned by French and Spanish firms. Three different processes are currently used to make sparkling wines in the United States: *la méthode champenoise*, the "transfer" method, and the Charmat (or bulk) process. Real French Champagne is made *only* by *la méthode champenoise*. In the United States, the process by which a sparkling wine is made must appear on the label, a clue at least for knowledgeable consumers as to what to expect when they open the bottle.

La Méthode Champenoise

The best American (or any other) sparkling wines are made by *la méthode champenoise*, though acceptable to very good sparkling wines can be made by the other processes as well, especially by the transfer process. Irrespective of the process that is used to make the wine "sparkling," it is essential to begin with a sound base wine, usually a dry white table wine. Thus, all sparkling wines go through two separate wine-making processes—one to make the base wine, the other to add the bubbles.

In France's Champagne district, only three grapes can legally be used for the production of sparkling wines: Chardonnay, Pinot Noir, and Pinot Meunier. In the United States, however, numerous grapes are used, depending on cost and the type of market that the sparkling wine is being produced for; there are no legal constraints. Don't expect those $2.00 bargain sparklers to be full of Chardonnay or Pinot Noir, however! Among American producers who use *la méthode champenoise*, though, most now use mainly the classic Champagne grapes, blended in various percentages.

Sparkling wines are made by adding measured quantities of yeast and sugar to the base wine, thus inducing a carefully controlled "secondary" fermentation within a closed container. The carbon dioxide released by this new fermentation is "trapped" in the container rather

than being released into the atmosphere, as it was during the initial fermentation. It must be controlled in order to keep the pressure in the bottle from exceeding what the bottle can hold and exploding the container (a common occurrence in earlier days, before the process was quite so well understood).

In *la méthode champenoise*, the new mixture is placed in strong bottles, which are then sealed and laid horizontally. Secondary fermentation takes place in the bottle, wherein the resulting carbon dioxide is trapped and forms the bubbles. At this point, the pressure inside of the bottle is typically in the range of 80–90 pounds or more per square inch (compared to the normal auto tire pressure of around 32 pounds per square inch), enough to destroy a weak bottle and certainly enough to blow the cork out when you open it (which is why warning labels often appear on bottles and why people should be careful in extracting champagne corks). After the secondary fermentation is complete, the yeast cells die (for lack of food) and settle out. The bottles are then aged for anywhere from one to four or more years.

The next step in the process is "riddling," which, by turning and slightly jarring each bottle, then gradually turning each bottle nearly upside down, places the remaining yeast sediment on the cap. Riddling in France is still often done by hand, but in the United States automatic riddling racks are the general rule. Once the riddling is finished, it is time for "disgorging," or removing the accumulated sediment. The neck of each bottle is frozen, and the cap, with its sediment neatly attached, is removed, leaving a clear sparkling wine.

Before recorking, a *dosage* is added to each bottle. The *dosage* is a mixture, usually of sugar and wine, that brings the level of each bottle back to the top and, at the same time, adjusts the final sweetness of the wine. In the United States, wines made by this complex and laborious method will indicate that fact on the label, in language such as "Naturally fermented in this bottle" or "*Méthode Champenoise*," or both. When the former is chosen, the emphasis is on *this*, as will be apparent shortly—in this process the wine never leaves the original bottle.

The Transfer Process

The transfer process begins the same way as *la méthode champenoise*. However, the product, after secondary fermentation is complete, is then emptied from the original bottles into a pressurized tank, from which it is filtered and rebottled. The label will say "Naturally fermented in the bottle." The subtle labeling difference, *this* or *the*, is hardly apparent to the typical supermarket shopper looking for a bottle of bubbly, yet it is a clear and legal delimitation between the two production methods!

The Charmat, or Bulk, Process

The Charmat, or bulk, process, is the least expensive of the three, and on the label it must state that the Charmat, or bulk process, was used. As you might guess, most inexpensive sparkling wines are made by this process. The base wine (usually from common and inexpensive grapes), with more yeast and sugar added, goes into large tanks. The wine undergoes a secondary fermentation in these tanks, after which it is filtered and bottled. This is what most Americans have experienced as champagne, leaving them with something to look forward to.

Fortified Wines

Fortified wines are made by adding ethanol distilled from wine. Broadly speaking, as Amerine and Singleton (1977:150) noted, fortified wines "may be defined as that group of wines produced by the addition of wine spirits." The best-known American examples are generically labeled "port" and "sherry"—using, as with Champagne, the names of European counterparts; ports, as well as most sherries made in the United States, are dessert wines.

Much heftier than table wines, fortified wines normally have an alcohol content of between 18 and 21 percent, though a few are somewhat lower. As esteemed French enologist Emile Peynaud (1984:226) noted about dessert versions, "They are sweet because by means of halting the fermentation with added alcohol they retain a large part of the sugar from the original must." The addition of alcohol stops fermentation because most strains of yeast cannot survive at alcohol levels above about 15 percent. However, according to Amerine and Singleton (1977:150), "The level of 17 to 18 percent alcohol is about the minimum that will reliably make the wine microbiologically stable." It's better to be safe than sorry: Wine makers don't want any unwanted surprises such as renewed fermentation after the wine has been bottled.

Fortified dessert wines require ripe grapes with good sugar content, usually 23 to 25 degrees Brix, so the bulk of these wines in America are made from grapes grown in California's Great Central Valley, where long, hot summers can fully ripen them. However, a recent trend toward smaller producers making premium dessert wines has emerged. Most of them are making ports, and many of them are located outside the Great Central Valley. At one time, southern California's Cucamonga Valley was important for the production of fortified dessert wines as well, though today there are few vines and even fewer wineries there.

The distilled spirits used to make fortified wines must be at least 185 proof and they must have been made from grapes; they are typically neutral in flavor. Fortified wines once composed a much higher percentage of America's wine production than they do today, and they have undoubtedly contributed to the national stereotype of "winos."

Fortified wines may be made from either white or red grapes. Examples of white fortified wines include White Port, Angelica, and Muscatel. Port, often labeled as ruby or tawny, is the primary red wine in the dessert category.

Sherries are somewhat different from ports or other fortified wines. Unlike most wines of any kind, sherries require a slow oxi-

ESTATE BOTTLED IN OUR CELLARS BY

FICKLIN

Vineyards

MADERA
CALIFORNIA

California

PORT

OF MADERA
ALCOHOL, 18.5% BY VOL.

dation, or "maderization," generally either from long-term aging or from artificial "baking." An alternative way of producing sherry is to use a very special type of yeast, the *flor* yeast. Most of America's sherries are not produced by the *flor* process, however. Along with sherries, other American wines that are produced in similar fashion include Madeira, Malaga, and Marsala.

Most generic (e.g., port or sherry) fortified wines differ from the original wines from which they have derived their names, though premium producers here often make products that are more similar to the originals. Thus, American ports and sherries, for example, are typically different from their Portuguese and Spanish counterparts. These disparities are a result not only of geographic differences but also of different grape varieties and different wine-making techniques. Of course, this does not mean that you won't find good American versions of ports and sherries, because indeed you can; rather, the question in some minds is whether to use the generic names. It seems unlikely that we'll give up the practice.

Modern Changes in Wine Making

Recent technological advances allow wine makers to control the wine-making process more carefully than ever before. The use of stainless steel containers for fermentation, along with careful temperature control during fermentation, has revolutionized the making of white wines. Quality control during the harvest and crushing has improved also, with such innovations as night harvesting to avoid high grape temperatures and field crushing to more quickly separate juice from grapes. Improved techniques have been developed for sulfur dioxide determination, for monitoring sugar levels during fermentation, for stabilizing wines, and for controlling malo-lactic fermentation (a secondary process in which malic acid in the new wine is converted into lactic acid). Various chromatographic methods have been developed for determining in detail the composition of grapes and wines, though they have not yet replaced human tasting and sensory evaluation.

Within the American wine industry, considerable experimenting still goes on with such things as fermentation in different types of containers, aging in a variety of barrels, and blending. Some fermentation of white wines is done in oak, for example, especially with Chardonnay and occasionally with Sauvignon Blanc. Aging experiments involve evaluating the use of various barrel sizes and types—the benefits of large or small barrels, for example, or the various effects of American oak, French oak, German oak, or Yugoslavian oak—and investigating the results of various lengths of times wines spend in oak. Sometimes wines are even blended after aging in different kinds of oak in order to "marry" their various desirable characteristics. Too many wineries in this country, however, persist in overemphasizing oak at the expense of the fruit's own subtle characteristics. As Leonard Bernstein commented in his *The Official Guide to Wine Snobbery* (1982:140): "A California Chardonnay will be poured and tasted, and someone at the table will comment, 'Too much oak.' A solemn nodding of the heads will follow as though nine justices of the Supreme Court had pronounced unanimously on a First Amendment violation." Oregon, Washington, Texas, and Long Island Chardonnays don't seem quite so susceptible to either overoaking or ridicule.

Today's wine making is a thoroughly scientific endeavor, though perhaps some of the magic has been lost, some of the romanticism destroyed. Nonetheless, considerable gains have been made as well, and the reliable production of sound, drinkable table wines has, for the most part, replaced hit-or-miss tactics in the American wine industry, much to the benefit of America's wine consumers. Colorful characters such as Bonny Doon's Randall Graham remain, however, and the wineries are not yet producing a homogeneous product without noticeable differences. Even so, many consumers do lament the current emphasis on Chardonnay and Cabernet Sauvignon at the expense of so many other interesting grape varieties. This is not to say, of course, that all American wineries have reached the same standards of production; a few are still fermenting in odd containers, aging wine in inexpensive plastic vats, and struggling at the margin of modern wine growing.

Subjective decisions about when to pick, how to ferment, how long to age, which oak to use, and a host of other things still yield wines that are personalized by their makers. What has diminished considerably in recent decades is the probability that you will find an absolutely terrible bottle of American wine.

Locations of American Wineries

Figure 4.1 provides a brief overview of the largest wineries (which often have more than one facility) in the United States. The primary determinant of the location of a winery is proximity to grape supplies, so that grapes can be moved from field to crusher in a minimum of time in order to minimize spoilage, juice loss, evaporation, oxidation, and the growth of undesirable yeasts and other organisms. Thus, wineries

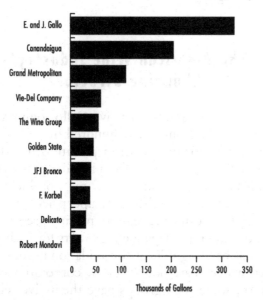

FIGURE 4.1   Largest Wineries in the United States: Total Storage Capacity, 1995
Many of these wineries have more than one facility. *Source:* Data are from *Wines and Vines* 76, 7 (July 1995), p. 51.

are located primarily in or very close to the major grape-growing regions. A majority of wineries own vineyards, though they may also buy grapes from other vineyards. Grapes and wines move around wine-growing regions in sometimes mysterious patterns; for example, many unmarked (and sometimes marked) vehicles that pass by on California freeways are transporting grapes and wine in tanker trucks.

There are exceptions to the locational pattern of wineries described here, of course, though they are neither widespread nor numerous. One example is the location of wineries in urban areas in the San Francisco Bay Area; East Bay cities such as Oakland and Berkeley have wineries without vineyards. Wineries locating in these areas has been a result of several factors, including relatively low rents in buildings that once had other uses (old warehouses, for example) and proximity to the large Bay Area market for wine sales. However, note that it makes sense for any winery that does not own a vineyard to locate in the Bay Area—a central location with relatively easy access to grapes from nearby Sonoma, Napa, Alameda, Santa Clara, and Monterey Counties. Grapes from these viticultural areas can be field crushed and trucked within minutes or at most a few hours to the Bay Area for processing.

The American Wine Industry's Changing Structure

Wineries range in size from the massive Gallo complex to small-scale operations that produce only a few hundred or thousand cases annually, and the vast majority remain in private ownership. However, modern wine-making equipment, from bladder presses to oak barrels, is expensive; insufficient capital has become a problem, especially for many small wineries. In addition, grape prices have been rising steeply in most American wine regions, putting pressure especially on wineries that depend on independent growers for their grapes.

One result has been significant consolidation in recent years, as undercapitalized wineries are bought up by corporations or investment groups; in turn, some such groups have themselves changed hands. Wine World Estates, known now as Beringer Wine Estates (and previously owned by Nestlé) is perhaps the best recent example. Now owned by Silverado Partners and Texas Pacific, Beringer Wine Estates currently includes Beringer, Napa Ridge, Chateau Souverain, Chateau St. Jean, and Meridian Vineyards among its wineries (with perhaps 9,000 acres of vines), is part-owner and sole distributor of Maison Deutz sparkling wines, and has an annual gross revenue of approximately $200 million. American spirits company Brown-Forman provides another example of winery consolidation, owning Korbel, Fetzer, Bel Arbor, and Jekel—all California wineries with excellent reputations—along with Jack Daniel's and several other whiskey labels, as well as Jack Daniel's 1866 Classic Amber Lager, one of many recent entrants in the microbrewery industry. Other spirits companies that have winery holdings include Seagram Classics (with Sterling, Mumm, and Monterey Vineyard), Heublein (in turn owned by English Grand Metropolitan PLC and including Beaulieu and Heublein Wines), and Stimson Lane (with Chateau Ste. Michelle, Columbia Crest, Villa Mt. Eden, Whidbey's, Snoqualmie, and Conn Creek).

Although Gallo is still the nation's largest winery, Canandaigua is now second and is steadily gaining ground, in part by buying up other wineries and their properties—including Paul Masson, Taylor, Almaden, and Inglenook. Other wine labels under the Canandaigua umbrella include Cook's (American Champagne and a new line of varietal wines), Cribari, Manischewitz, J. Rogét, Dunnewood, and Marcus James (imported from Brazil). In addition, other wineries are busy buying up wineries and properties: Robert Mondavi, for example, also owns Vichon, Mondavi-Woodbridge, and Byron (along with one-half of Opus One); Kendall-Jackson owns Vinwood Cellars, Cambria, J. Stonestreet, R. Pepi, La Crema, Lakewood, and Edmeades; Windsor Vineyards also owns Rodney Strong and Great River Winery; the Chalone Wine Group now includes Acacia, Edna Valley, Carmenet, and Canoe Ridge, along with Chalone Vineyard; and the Wine Alliance includes Clos du Bois, Callaway, Atlas Peak, and William Hill.

Many newer wineries are heavily in debt, so consolidation in some form (joining with other wineries, for example) is one way to cut costs. For example, corks and bottles can be bought in larger quantities with lower per unit costs, equipment can be shared, marketing and advertising costs can be combined, and some wineries even share wine makers. Without doubt, a shakeout has been taking place in America's wine industry, and it is a long way from being over.

5

𝒲ine Regions and Wine Labels

$\mathscr{P}$REVIOUS CHAPTERS HAVE PROVIDED information about the history of viticulture, major cultivars in use today across the nation, vines and their environmental requirements, and the rudiments of modern wine making and its technology. These various factors have come together in selected locations to create and shape modern American wine regions as we approach the new millennium.

American Wine Regions

Although still dominated numerically by California's vast vineyards and multitude of wineries, the number and variety of American wine regions have grown considerably during the past three decades, bolstered alike by a pioneering spirit among new winegrowers and local boosterism.

Viticultural regions emerge when there is a fortuitous combination of natural and social conditions that favor their development. Without environmental circumstances that provide nurturing environments for grapes, other crops (or perhaps no crops) will dominate. However, areas that are capable of producing wine grapes can also support numerous other crops, including prunes, apples, cherries, almonds, walnuts, olives, and a host of field and row crops. Wine grapes will be grown (and wines produced) mainly where and when they are able to compete with other crops in maximizing the net return per acre to farmers.

Beginning in the 1960s, Americans' interest in wines increased perceptibly, and the nation's wine makers began to stir as well, awakening to shifting demands and tastes among consumers. Coupled with this revitalized consumer interest was the appearance of some encouraging technological improvements, especially the new stainless steel fermentation tanks that were being introduced and the use of temperature-controlled fermentation techniques. First in California and New York, long the strongholds of the nation's wine industry at that time, but soon in many pristine areas as well, new vineyards were being planted and modern wineries, small and large, were being built.

In the early 1970s, a study published by Bank of America (1973) projected a growing demand for wines among American consumers. That study served as the basis for encouraging new financing in the industry, allowing growers and wine makers to move forward with their plans. Another source of capital was wealthy professionals who thought that owning a winery, some vineyards, and maybe even a home in "wine country" would be rewarding. For example, a number of doctors, dentists, celebrities, and successful entrepreneurs provided an impetus to expansion of the wine industry in California and elsewhere from the mid-1970s onward.

Until the 1960s, sweet and fortified wines dominated the nation's wine production. However, by the 1970s a serious shift toward table wine production was well under way. A further boost for the reviving American wine industry came in 1976, at the Bicentennial Tasting in Paris. There a panel of French wine experts tasted numerous wines in a blind tasting, featuring both French Bordeaux wines in competition with California Cabernet Sauvignons and French White Burgundies in competition with California Chardonnays. Surprise winners were two California wines, the 1973 Cabernet Sauvignon from Stags Leap Wine Cellars and the 1973 Chardonnay from Chateau Montelena—the world-class status of California wines had been confirmed.

During the expansion of the American wine industry since the 1970s, commercial viticulture and wine production have spread into virtually every state in the union that could find niches, however small, within which to grow wine grapes. Encouraged by both consumer acceptance of their wines and a series of "farm winery" laws that made the making and selling of wine easier in many states, commercial wine making quickly spread to states as diverse as Arizona and New Mexico, on the one hand, and Virginia, Massachusetts, and New Hampshire, on the other. Only a half-dozen states have no commercial wineries. Maps 1 and 2 show, respectively, the current distribution of wine grapes harvested and the commercial wineries in the United States.

Viticultural areas have emerged in the 1990s as a geographic expression of this fortunate combination of events, aided by official recognition from the U.S. Bureau of Alcohol, Tobacco, and Firearms, which oversees everything from the establishment of official viticultural areas to the approval of labels that adorn bottles of the final product of the wine maker's art. At the same time, official recognition of viticultural areas was supplemented to a considerable degree by local boosterism—local wineries encouraged visitors to come, taste wines, and perhaps even picnic at the wineries (many have excellent picnic facilities, including delicatessens at some). In turn, local communities began to encourage tourists to visit areas where at least a few

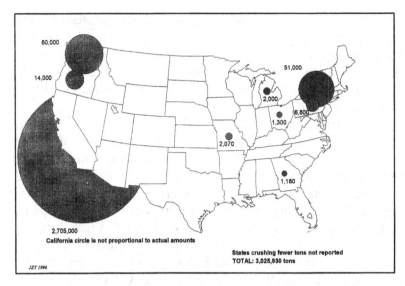

MAP 1   Wine Grapes Harvested in Tons in the United States, 1995
The tonnage amounts are based on preliminary estimates made by the U.S. Department of Agriculture.

wineries were beginning to operate. Local fairs and festivals were organized, with wine judgings, harvest celebrations, hot-air balloon rides, outdoor concerts, and a host of other events reinforcing the positive experiences to be had as a part of visits to wine country. From Temecula in southern California to Cutchogue on eastern Long Island, New York, landscapes have been changed in varying degrees by either the introduction or expansion of wine making during the past three decades.

Across America, more people are scheduling visits to wine country, as is attested to, for example, by the American Automobile Association's publication of *California Winery Tours* (Automobile Club of Southern California 1995), its sponsorship and presentation of *Guide to the Best Wineries of North America* (Gayot 1993), and special notes about wine regions in many of its tour books. Further evidence of local interest in wine regions can be seen in promotions by local chambers of commerce, wine columns and wine news in local newspapers and magazines, and discussions of wine country travel in consumer-oriented food, wine, and travel magazines. Once viewed by most people as being limited to California's Napa Valley (and perhaps New York's Finger Lakes region), wine country of some sort is now avail-

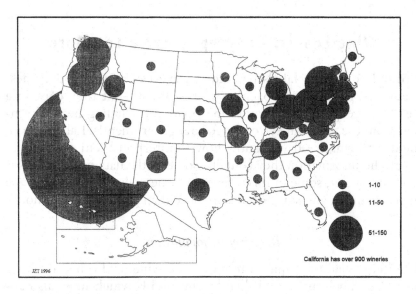

MAP 2   U.S. Wineries by State, 1996
Source: Data from U.S. Department of the Treasury, *Bonded Wineries and Bonded Wine Cellars Authorized to Operate* (Washington, DC: U.S. Government Printing Office, 1990).

able to travelers in a majority of states. Wine-related landscapes are attracting a growing number of visitors, many of whom contribute to local economies not just by purchasing wine but also by eating in local restaurants, staying in local hostelries, and buying local gasoline and other necessities and sundries. Wine country visits, then, are being promoted in some regions not out of local provincialism alone but also out of perceived benefits to local economies, many of which have languished for decades as rural folk moved to cities and left behind landscapes that offered little or no appeal to either investors or visitors. The health benefits of wine have been lauded during the 1990s, though this was so far earlier, even if the praise had a less scientific basis. There may also be some psychological benefits to be derived from trips to local wine regions, wherever you find them.

These regions and the winescapes within them are considered in detail in the next chapter. Before that, a brief foray into the political geography of viticulture is necessary. Because wine is an alcoholic beverage, wine labels are carefully controlled in many ways, and we can learn much from them about wines, wine regions, and social concerns.

Wine Labels: Geography and a Lot More

Not only can wine labels be colorful, eye-catching, and even aesthetic—sometimes approaching works of art—but they contain a wealth of geographic and other information (including health warnings from the watchful eye of the federal government!). Labels can be highly informative if you know how to interpret them and how to weigh the information they contain. Some of the data on the labels is required by law, some of it is offered by wine makers to help you, the final consumer, and some of it is no more than good public relations.

Required Information

The items that are required to appear on wine (and other alcoholic beverage) labels are established by the BATF, which must also approve each and every label that goes on a bottle of American wine. At minimum, each wine label must include a brand name, identification of the type of wine that is in the bottle, its alcohol content, information about processing and bottling, sulfite content and other warnings, and an appellation of origin—a purely geographic concept that can help consumers make informed choices.

The Brand Name

A brand name must appear on every wine label. Most of the time, this is the name of the registered and bonded winery itself, though occasionally second labels (which also must be legally approved) are used by wineries for certain products. Second labels are often used when a wine is not quite up to the standard of the main label but not so bad that it cannot be sold; they are also used to distinguish between lines of wine from a particular winery. Despite the importance of regional appellations and other information, most consumers

SAKONNET

ESTATE BOTTLED

Pinot Noir

SOUTHEASTERN NEW ENGLAND

GROWN, PRODUCED AND BOTTLED BY
SAKONNET VINEYARDS, LITTLE COMPTON, RI
ALCOHOL 11.1% BY VOLUME

1992

are still guided primarily (and, I might add, correctly, for the most part) by the brand name on the bottle; find producers that you like and you don't have to worry much about the more complicated issues associated with appellations and other details. Nonetheless, there are now hundreds of different second labels in the United States, and knowing something about them can often lead budget-minded consumers to good buys.

The Type of Wine in the Bottle

Each label must identify the type of wine that is contained in the bottle, such as table wine, dessert wine, fortified wine, or sparkling wine. The most important ways of identifying table wines are as (1) varietal wines, (2) generic wines, and (3) proprietary wines.

Varietal Wines. Varietal wines are designated by the grape variety from which they are made, such as Cabernet Sauvignon, Sangiovese, Zinfandel, or Chardonnay. In order to use the varietal name on the label, however, at least 75 percent of the wine must have been made from that grape (which still means that a Chardonnay, for example, could include up to 25 percent of wine made from other grapes). On the one hand, the 75 percent minimum level is well above the 51 percent level for varietal wines that prevailed until 1978. On the other hand, Jess Jackson (owner of Kendall-Jackson Vineyards) began actively campaigning in the mid-1990s to raise the current 75 percent minimum to 85 percent. Whether this is a good idea or not depends on your perspective—in some ways too much has already been made of varietal labeling, and blending wines made from different grapes is becoming more common again. There is nothing essentially better about a varietal wine compared to a blend; the quality of what is in the bottle depends on grapes, weather, and human skills.

Generic Wines. Generic wines are typically labeled with broad regional names (most often those of European wine regions, much to the chagrin of foreign producers) such as Burgundy, Chablis, and Chianti. The implication, of course, is that there is some similarity between what is in the bottle and what you might find in a bottle of wine imported from that European wine region. Nothing could stray much further from the truth. There are no restrictions on the types of grapes that are contained in these generic wines; most are blends of different grapes, usually from warm growing regions. A California or New York Chablis, for example, is likely to have lots of French Colombard or even some hybrid juice in it (and maybe even some Thompson Seedless in some cases); it is unlikely to contain any Chardonnay, even though that would be the only ingredient in a "real" French Chablis. Fortunately, American wine makers have moved away from most generic labels, even for "jug" wines, those mass-produced, inexpensive, but often quite pleasant and well-made products from producers located mainly in California's Great Central Valley and in viticultural pockets from Washington to New York and Virginia.

Proprietary Wines.   Proprietary wines, those labeled with names that are "invented" by the winery, are becoming more common (both at the low- and high-price levels, as noted already concerning Meritage wines), perhaps because American wineries are now more interested in distancing themselves from the names of European wine regions and less interested in invidious comparisons. Many (perhaps most) proprietary names have no meaning at all; some say no more

than "Red Table Wine," though others are often quite creative— Bonny Doon's Le Cigare Volant, for example, or Hop Kiln's Thousand Flowers. Other examples include the following: Horton Vineyards offers Côtes d'Orange; Heitz Wine Cellars produces Ryan's Red, named after Joseph Heitz's first grandson; Boeger Winery, located in Placerville, in California's "Gold Country," produces Hangtown Red, named after the once-common nickname of Placerville as "Hangtown"; Washington's Bainbridge Island Winery makes an aptly named Ferryboat White; Willamette Valley Vineyards produces Ore-

gon Edelweiss; Oakencroft makes Countryside White; and Leeward Winery offers Coral, a blush wine. Occasionally the line between generic and proprietary becomes blurred, as with Wiederkehr's Alpine Chablis.

Elsewhere on the proprietary label you will usually (but not always!) find some indication of what grapes the wine in the bottle was actually made from. For example, the back label might indicate that a proprietary red wine is made from 50 percent Petite Sirah and 50 percent Cabernet Sauvignon (so it could legally be called neither Petite Sirah nor Cabernet Sauvignon because of the 75 percent rule).

Nor are all proprietary labels used for inexpensive table wines anymore. Many are used to designate premium and even ultrapremium wines: Meritage blends (red or white, almost exclusively using grape varieties that would be used in France's Bordeaux blends), Italian blends (which may also include some Zinfandel or Cabernet Sauvignon), and Rhône blends are good examples. They often do not contain a sufficient percentage of wine from one cultivar to be labeled as a varietal wine, yet they have been made with expensive grapes and considerable care. Insignia from Joseph Phelps is a good example, as is Opus One, the famous Mondavi-Rothschild wine, or Trilogy from Flora Springs. Somewhat more satirically, Gundlach-Bundschu offers a wine that it calls Bearitage, whereas Gold Hill makes a

blend that it simply labels as Meritage Red Table Wine. Sometimes such blends are indicated differently, as with Prince Michel's Merlot Cabernet (a blend of 66 percent Merlot and 34 percent Cabernet Sauvignon). Similarly, Arbor Crest Wine Cellars produces Cabernet-Merlot, which, surprisingly, is a blend identical to Prince Michel's in its proportions of Merlot and Cabernet Sauvignon.

Alcohol Content

A wine's alcohol content must also be indicated, though this can be accomplished in different ways, depending on the type of wine and the desire of the wine maker. For wines that are over 14 percent alcohol, the actual alcohol percentage must be given, accurate to within 1 percent. Dessert wines that have been fortified with brandy can be labeled by type (as port, for example), with a range shown for alcohol content, to a maximum of 24 percent.

Most table wines are under 14 percent in alcohol; their alcohol content can be shown in one of two ways. The percentage of alcohol, accurate to within 1.5 percent, can be used *or* they can simply use the term "table wine" or a modification such as "red table wine" or "white table wine."

Processing and Bottling Information

The name and geographic location of the bottler must appear on the label, and it can do so in different ways, depending on the role that the bottler had in making and cellaring the wine. Note, however, that the place of bottling is not necessarily a guide to where the grapes from which the wine was made were actually grown—more about that in a subsequent section. The words "produced and bottled by" can be used *only* if the bottler fermented and cellared at least 75 percent of the wine in that bottle. In contrast to the above, the expression "made and bottled by" can be used if the bottler fermented at least 10 percent of the wine in the bottle. When neither of the above conditions are met, then other expressions, such as "vinified and bottled by" or "cellared and bottled by," can be used, as can simply "bottled by." Whereas the former suggests that the winery may have done something with the wine before bottling it, the latter means literally what it says, that the winery bottled a wine that was made and cellared by someone else. Although this may surprise some of you, perhaps even jolt your romantic image of the American wine industry, it is a common practice. Unlabeled tanker trucks move wines around and between wine regions in a swirl of patterns, and at harvest time, trucks may be seen heading off from vineyards in all directions.

If you reflect upon it, however, the practice makes considerable sense. A winery is obviously immobile (though grapes can be crushed in the field), so in order for it to take advantage of grapes from different growers and regions, it may very well have to transport them across designated viticultural boundaries. Well-known and respected growers and vineyards have grapes that are in demand. Grapes from these vineyards may be sold to many different wineries, and occasionally grapes, juice, and wine are even carried across state borders.

Required Warnings

The BATF requires that certain warnings must appear on labels. For example, a wine label must indicate the presence of sulfites in a wine (primarily because some people have an allergic reaction to them). Sulfur has been added to most newly harvested grapes used in commercial wines, so almost all commercial wines contain sulfites; a few organic wines do not include sulfites, however, and they can therefore drop this warning from their labels.

Another government warning label must be affixed. It includes two separate items: (1) a warning about possible birth defects that might result if a woman consumes wine during pregnancy (because of concerns about the possibility of fetal alcohol syndrome) and (2) a warning that the consumption of alcohol can impair your ability to drive a car or operate machinery and may cause health problems. Fortunately (for both the wine industry and wine consumers), in 1996 a more enlightened set of federal food guidelines was published that for the first time suggested that moderate consumption of alcohol may actually have health benefits. Perhaps in the coming millennium, wine labels will be allowed to suggest health benefits as well— but don't hold your breath. Those troubled over these warnings on wine labels can take some solace in the unappetizing required labeling for products that contain Olestra (a fat substitute manufactured by Procter and Gamble, which is none too happy with the requirement): That label must warn about the possibility of "abdominal cramping and loose stools."

Appellations of Origin

Each wine label must carry an indication of the wine's geographic origin; it must identify, with varying degrees of geographical specificity, the region where the grapes were grown. As a result, wine labels offer valuable lessons in the regional geography of viticulture. We find a geographic hierarchy on wine labels that ranges from American (usually used when the wine has been produced from grapes that were grown in two or more different states) to the levels of the state, the county, and the American Viticultural Area (AVA).

The broadest geographic appellation, of course, is simply American. It can be used when the wine was made from grapes grown in more than one state (but totally within the United States) *and* when the maker cannot, or chooses not to, use a multistate designation. Cook's American Champagne, for example, carries an American appellation, though it is bottled in Woodbridge, California.

A multistate appellation is allowed if the following conditions are met: (1) a maximum of three states can be used *and* they *must* be contiguous, (2) the grapes must be grown within the states designated *and* the percentage of grapes from each state must appear on the label, (3) with only very minor exceptions (finishing, blending), the wine must have been completely processed within the states named on the label, and (4) the wine must conform to the wine laws of each of the states listed in the appellation.

A state can be used as an appellation of origin, in which case at least 75 percent of the wine must be made from grapes grown within the state. If a wine's origin is simply California, then 100 percent of the grapes for that wine must have been grown in California (in this case California law is stricter than federal law, which requires only 75 percent of the grapes to be grown in the state in order to use the state name as the region of origin). Oregon also has wine laws that are stricter than the federal ones.

If a wine is labeled with a county region of origin—San Luis Obispo County, for example—then at least 75 percent of the grapes used to make that wine must have been grown within that county. Multiple county origins, such as Sonoma *and* Mendocino, may be used on a wine label also, but the percentage of grapes from each county must appear on the label. Multicounty designations are not common because keeping track of the percentages of grapes from each county may be more trouble than it is worth (so the wine may simply be labeled "California" in that case).

The smallest officially recognized geographical unit that can be used as the appellation of origin on a wine label is the American Viticultural Area. If an AVA, such as Los Carneros or Texas Hill Country, is used as the appellation of origin, then at least 85 percent of the grapes must have been grown within that AVA.

With the official designation of American Viticultural Areas in the early 1980s came a new legal definition of the term "estate bottled" as well. Although it is optional on a wine label, the term can now be used *only* if a winery is located within an AVA. Furthermore, the wine must have been made entirely within the winery from vineyards that are either owned by, or are at least completely under the control of (usually by means of long-term contracts), that winery *and* are within the AVA.

Among people who profess a serious interest in wine, most would agree that knowledge about a wine's precise geographical origin is important. As geographer Harm de Blij (1988:37) nicely put it, "In the relationship between producer and consumer, the wine map is of incalculable value." The classification of wine regions has long been important, as the wine laws of France, Italy, and Germany most assuredly illustrate. However, until recently in the United States, we have been satisfied with defining areas of origins for wines using only state or county boundaries, despite considerable variations in soils and microclimatic conditions within those broad political units.

In 1978, the BATF decided to recognize a more detailed system of areas or appellations of origin for wines produced in the United States; thus the viticultural area concept was born. On January 1, 1983, new regulations concerning the appellation of origin on American wine labels became effective, and recognition was given to officially approved viticultural areas, now called American Viticultural Areas. These appellations can also be used for advertising. As mentioned already, if a viticultural area designation is used, then at least 85 percent of the wine must be made from grapes that were grown in that AVA.

Viticultural areas must be approved by the BATF, with input from wine industry members, viticulturists, consumers, and others who may have an interest, often monetary or political, in the process. Curiously, professional geographers, despite their interest in establishing regions of every sort, have seldom been consulted.

The procedure begins with the submission of a petition to establish a new AVA. This petition is ordinarily sent in by one or more wineries or one or more grape growers who seek to establish the area as unique for viticulture. The petition must present at least the following information:

1. Evidence that the name of the desired viticultural area is at least locally known as referring to the area specified in the proposed boundaries
2. Current or historic evidence that the boundaries of the viticultural area are indeed those that are specified in the petition
3. Evidence relating to the geographic features, such as climate, topographic expression, elevation, and soils, that differentiate the proposed viticultural area from the surrounding region
4. A copy of the appropriate U.S. Geological Survey map (or maps) showing precisely how the proposed viticultural area boundaries are to be drawn

The identification and establishment of American Viticultural Areas is clearly a problem in applied regional geography, though to my knowledge only one professional geographer, William Crowley at Sonoma State College, has been directly involved in any of the petition designs.

Once the petition is submitted, it is then reviewed by the BATF, which also seeks public comment. In some cases, public hearings may be convened if significant differences in opinion must be resolved. Subsequently, a U.S. Department of the Treasury decision is issued establishing the official AVA. An up-to-date list of AVAs is available on the Internet. Unfortunately, maps of the AVAs are not available for those established since about 1986 because of budget limitations at the BATF.

Identifying these areas helps to provide consumers with more specific geographical information about a wine's origin and allows wineries to better designate where they grow or buy grapes for their wines. However, the BATF has been clear in communicating that the establishment of an AVA is *not* a guarantee of wine quality. Nonetheless, AVAs may ultimately help improve the image of U.S. wines in foreign markets; American wine exports have been growing steadily in recent years, especially to Canada, England, and Japan.

However, there are some significant differences between the American appellation system and its counterparts in nations such as France and Italy. Within most European "appellations" (legally delimited geographic regions), the administrative bodies that oversee them tend to control much more than just the geographic outlines for viticulture; they exert considerable control over viticultural and enological practices within those boundaries as well. For example, within French appellations there are controls over the cultivars that can be used for wine making, the planting and pruning of vines, the yields per hectare that can be produced by growers, and sometimes even the minimum alcohol levels of regional wines. Such rules establish clear associations between regional viticultural practices and the wines that are produced there. That is not the case with American Viticultural Areas.

French Champagne, as noted already, can be produced only in the Champagne region of France and only from some combination of one or more of the legally allowed grapes: Chardonnay, Pinot Noir, and Pinot Meunier. Furthermore, French Champagnes can only be produced by *la méthode champenoise* (also legally recognized within the European Union as *la méthode classique*, *la méthode traditionelle*, and *la méthode traditionelle classique*). In the United States, there are restrictions neither on the cultivars that can be included in American sparkling wines nor on the process used to make them. However, as was made clear earlier, American wine makers must indicate on the label the process that was employed, and the best of them use mainly, if not exclusively, Chardonnay and Pinot Noir grapes. Similar comparisons can be made between most American table wines and those produced in Burgundy, Bordeaux, or even Tuscany.

Map 3 shows the distribution of American Viticultural Areas in the United States as of the middle of 1996. Subsequent petitions are likely, and still more AVAs can be expected. Although it may have been a step in the right direction to establish AVAs, the viticultural area concept is not without its problems and controversies. Map 3 shows the geographic distribution of AVAs, but it does not show their relative size differences. And AVAs do vary considerably in size. There are large multicounty ones such as the Texas Hill Country AVA, the largest of all, covering 15,000 square miles and exceeding the size of any one of the nation's nine smallest states; three other AVAs exceed 5,000 square miles each. In contrast, other AVAs are small and relatively unknown. Among the smallest AVAs are Isle St. George, McDowell Valley, Cayuga Lake, and Cole Ranch (the na-

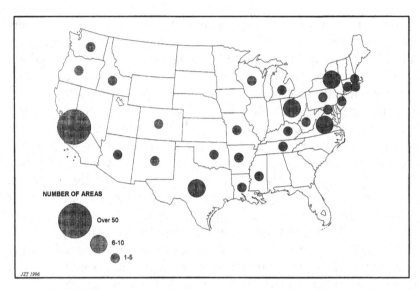

MAP 3   Viticultural Areas in the United States, 1996
Source: Data from U.S. Department of the Treasury, "U.S. Wine Appellations," 1996. World Wide Web.

tion's smallest AVA so far); furthermore, some small AVAs have only one winery, and at least one, the Potter Valley AVA in California, has no wineries at all. Other AVAs are very well known, for example, the Napa and Sonoma Valleys in California, Lancaster Valley in Pennsylvania, New York's Hudson River region, and Washington's Walla Walla Valley. So much variance among AVAs probably leads to more confusion than illumination for the average consumer.

Another problem that results in consumer confusion was described by geographer William Crowley as "nesting." This occurs when a broad AVA, such as California's North Coast, includes smaller AVAs, such as Los Carneros (which overlaps two different counties and AVAs) or Napa Valley (which in turn includes still smaller AVAs such as Rutherford, Oakville, and Stags Leap), completely within its boundaries. As grape varieties are more carefully matched with local microclimates and soils, the smaller AVAs may become increasingly important, but this will only happen in AVAs that can truly produce wines with distinctive characteristics that are derived from that AVA's unique viticultural environment or *terroir*. The association of California's Los Carneros AVA with fine Pinot Noir and Chardonnay is an excellent example. Larger viticultural areas are more likely to include a wider range of variations in pedology and microclimates and even mesoclimates (Regions I, II, and III are all represented in the Napa Valley, for example).

A different problem involves separating a winery's location from the region of origin of its grapes. Just because a winery is located in a particular AVA does not mean that all (or for that matter any) of its wines are made from grapes grown within that region. For example, the Sutter Home Winery is located in the Napa Valley AVA, but it makes a Zinfandel from grapes that are grown in Amador County. Similarly, the Los Carneros AVA appears on a wine made by Geyser Peak Winery in northern Sonoma County. And these two are hardly isolated examples. Rather, the practice of shipping grapes between viticultural areas is a common one, and it is likely to continue. Nothing is wrong with such practices, of course, but consumers need to be aware of them and to read the labels—and thanks to the BATF the information is all there for those willing to spend a little time learning what it all means.

Names of AVAs can themselves be confusing at times. In California, for example, there are two Green Valley AVAs, so each must include a county designation in order to differentiate between them (Solano—Green Valley AVA and Sonoma—Green Valley AVA). Similarly, there are two Shenandoah Valley AVAs, one in Virginia and one in California (the latter must be designated California Shenandoah Valley in order to distinguish it from *the* Shenandoah Valley). Although slightly different, the Escondido Valley AVA (Texas) could easily be confused with the San Pasqual Valley AVA (California), which is immediately adjacent to the town of Escondido in San Diego County.

One smaller geographical unit occasionally appears on labels, though by itself it is not truly an appellation of origin, and this describes the actual vineyard in which the grapes were grown, such as the Bien Nacido Vineyard in the Santa Maria Valley AVA or the Sangiacomo Vineyard in the Los Carneros AVA. If a vineyard designation is used, then at least 95 percent of the wine must have been made from grapes grown in that vineyard. Such designations are becoming more common. Other popular examples include Martha's Vineyard in the Napa Valley, Winery Lake Vineyard in the Los Carneros district, and Robert Young Vineyard in Sonoma County. Most single vineyards are either owned or controlled (usually with exclusive use) by one winery, though some individual vineyard owners sell grapes to two or more different wineries.

Optional Information

One optional item, the term "estate bottled," was discussed previously. If it is used, then it does have a precise legal meaning. However, a variety of other items, mostly without precise legal meanings, are also allowed to appear.

Labels often carry information that is useful to knowledgeable consumers, even though it is not required; figures for residual sugar and total acidity are good examples. "Private reserve," "select reserve," "limited release," and similar terms are occasionally used, though their meanings are imprecise and depend on the credibility of the winery (which is why brand names remain so important to consumers).

If vintage dates are used (as they are on most of the better table wines in the nation), then at least 95 percent of the wine must have been made from that year's harvest. It is less than 100 percent in order to allow for the topping-out of barrels, because over a period of time all barrels lose some of their contents to the atmosphere via evaporation (a loss sometimes known as the "angels' share").

Some wineries provide further information about the grapes, the harvest, and even the best foods to serve the wines with (even recipes have appeared on labels from time to time). Again, however, these may be imprecise statements, and it is up to the consumer to decide what is acceptable. Most wineries, though, make honest attempts to help consumers know what they are buying and to ensure maximum enjoyment of their products.

By now you can see the importance of carefully reading wine labels (especially if you are allergic to sulfites, pregnant, planning to drive or operate some heavy machinery, or all of the above). They are designed to convey a considerable amount of information, but that is of little value until you can interpret the label correctly and then evaluate the data that you have in a way that is meaningful to you. In turn, you can't do that until you have some idea whether the information is accurate. With the required items (e.g., region of origin, place of bottling, alcohol content), accuracy is well assured, but with other items you may be on your own. Labels also make it clear that an understanding of viticultural geography is important in order to fully appreciate what you are sipping.

6

$\mathscr{A}$merican Viticultural Landscapes

$\mathscr{V}$ITICULTURAL LANDSCAPES, or winescapes, are unique. The winsome combination of vineyards, wineries, and supporting activities necessary for modern wine production yields regions that offer sojourners and dwellers alike a certain charm—a warm ambience, a memorable experience of place—not found in most other agricultural landscapes. There is nothing more appealing or more alluring (not even "amber waves of grain" or any other crop, for that matter) than rows of neatly tended grapevines—vivid green and bursting with energy in summer, yellow-orange in fall, and starkly naked in winter.

Geography and Agricultural Landscapes

As geographer John Fraser Hart (1975:1) noted about rural landscapes more than two decades ago, "The countryside warrants far more attention than most of us have given it. The only proper way to learn about and understand it is to live in it, look at it, think about it, contemplate it, speculate about it, and ask questions about it."

According to Hart, rural landscapes are complex reflections of a combination of the topographic expression of the land surface itself, the kind of plants (both natural and domesticated) that cover that surface, and all types of structures that people have placed upon it to shape it. All these elements mold the landscape and give it order and meaning. For geographers, the summation of what we can see (and sometimes even hear, smell, and feel) in a place—a complex combination of its natural landscape and the visible evidence of human pursuits that occur within it—is conveyed by the term "cultural landscape." Agricultural landscapes, including viticultural landscapes, are cultural landscapes that are dominated by agricultural activities and the manifestation of those activities as they are expressed on the natural landscape. Put somewhat differently by geographers John Dickenson and John Salt (1982:184) within a viticultural context, "Ultimately, the geography of wine is an experience of place."

It is the essence of American winescapes that concerns us in this chapter. Seeking that essence, that sense of place that seems so palpable in viticultural areas, first requires some modest consideration of how geographers and others have viewed rural cultural landscapes more generally. Subsequently, viticultural landscapes themselves receive due consideration.

Neat rows of grapevines at Gold Hill Winery in the foothills of California's Sierra Nevada. The dark vines contrast sharply with the light grasses of nearby slopes.

Reading Agricultural Landscapes

Like books, landscapes of every sort can be read by those with suffi-cient knowledge and motivation, both for enjoyment and for enlight-enment. Long ago, the "natural" disappeared from most American landscapes, replaced to at least some degree with evidence of humans at work or at play. Even landscapes that at first might appear to be quite natural turn out to show at least subtle signs of alteration (plants that are not really native to an area are good examples). At the other end of the spectrum, of course, are our metropolitan behemoths, en-tirely artificial creations, perhaps best epitomized by Chicago, Los Angeles, and New York City. About that latter metropolis, eminent art historian Sir Kenneth Clark (1969:321) once wrote:

> Imagine an immensely speeded up movie of Manhattan Island during the last hundred years. It would look less like a work of man than like some tremendous natural upheaval. It's godless, it's brutal, it's violent— but one can't laugh it off, because in the energy, strength of will and mental grasp that have gone to make New York, materialism has tran-scended itself.

Although three-fourths of Americans now reside in metropolitan areas, many of them frequently seek respite in the vast open areas that can still be found from one side of the continent to the other (a night flight over the continent quickly demonstrates how much of it is still only lightly populated). Between the extremes of designated wilderness areas (with their appeal to limited numbers who seek serious solace) and megacities are our agricultural landscapes, a middle ground of working landscapes that include activities that range from the most extensive, such as the grazing of cattle and sheep, to the most intensive, such as raising strawberries near the metropolitan fringe. Americans have long found a certain comfort in spending time in these rural landscapes, even if only for a brief picnic and a quiet afternoon (an ideal use of wine country!).

The cultural landscapes of agricultural regions are shaped to a considerable extent by the nature of the dominant agricultural pursuits that are carried on. Over time, as such pursuits change, so does the essential nature of the agricultural landscape. For many years, geographers have chronicled such landscapes, their evolution and morphology, and from them we can learn what to look for as we move toward an understanding and appreciation of viticultural landscapes, which have remained relatively neglected by cultural geographers.

As John Fraser Hart (1975:67) has noted, "Any rural landscape is the product of a host of independent decisions made by the multitudes of individuals . . . who control and have controlled the individual pieces of land." What we see in agricultural landscapes, then, are expressions of such things as how the land has been divided into units (the rectangular survey system for most of the United States outside the original colonies and ranchos in parts of California and the Southwest are examples), the size of those units (from tiny plots to the megafarms of modern American agribusiness), ownership patterns, and the host of physical and socioeconomic factors that give rise to land-use decisions.

Anyone who has flown over the American Midwest has seen the neat rectangular fields, the proper east-west by north-south grid pattern of roads, and the geometric distribution of farmsteads and cities. They may also have discerned the strikingly different patterns created farther west, where water scarcity and economic necessity have given rise to center pivot irrigation systems that have modified the landscape, replacing the rectangular pattern of fields so common farther east with giant crop circles. In contrast to both of those patterns, the agricultural fields of the northeastern United States appear less regular, less neatly carved up, and the road patterns are not so rectangular either, often swirling around hills and through vales, reminding us of the constant influence that topography has had on how we use our landscapes.

Looking down from 30,000 feet, we can see the broad outlines created by dividing the American landscape according to a particular system or by impressing upon that system a particular type of farming. More subtle are the smaller features that constitute the landscape we see as we travel through it by train, car, bicycle, or on foot. Both views are useful in helping us understand what is going on, though slower modes of transportation, of course, allow more time for absorbing and reflecting upon the nature of landscapes and more opportunities to talk with those who have created them.

Geographers are inclined to consider both what can be seen in the landscape and what processes have led to the creation of that particular cultural landscape; most evolve out of long periods of changing land uses. Major elements of agricultural landscapes provide us with our "text": These include the major crop(s) and field patterns, buildings (such as barns, homes, equipment sheds, and wineries) and other material features (such as irrigation facilities, fences, and windrows), circulation systems (including roads, power lines, and telephone lines), and agricultural service centers (the villages and towns that act as local central places for the collection and distribution of goods and services).

Where agricultural landscapes are dominated by one or a few crops, distinctive patterns can easily be seen, shaped by the contours of the land, the physical and cultural needs of the crops, and the sizes and shapes of the fields. For example, in the American wheat belt, the vast wheat fields, the infrequent farmsteads and buildings, and the sparsely scattered towns (often with great silos standing silently along rail lines) create a landscape that is much different from the mixed farming region of the American Midwest, where an agricultural system based on corn, soybeans, and hogs demands more buildings, higher rural population densities, and more central places (to serve more people). Neither of these two very different landscapes could be confused with the cotton landscapes of the American South nor with the vast vineyard and orchard landscapes of California's Great Central

Valley. Even within the Central Valley, considerable differences exist between the agricultural landscapes of its eastern and western sides.

Buildings, often appealing in their own respect, in turn raise questions about their builders. Some buildings clearly reflect their functions (barns for livestock or silos for grain storage). Others may reflect the values and feelings of their owners; they may have been designed to be pleasing to look at as well as functional. Farmsteads may offer clues to the degree of success that farmers are experiencing. Even fences can tell us something about what we are experiencing in a landscape. Some are necessary, especially in livestock-producing regions; others may be purely decorative. "No Trespassing" signs in many areas provide visitors with what might seem to some a chilly reception, though farmers may simply not want people trampling down their crops (or stealing them).

Buildings designed for one purpose may gradually be adapted to others or may be left to disintegrate slowly as the weather and gravity take their toll year by year. With the shift to mechanized agriculture, for example, barns across America have fallen into disuse. Some have been saved by advertisers (some of you may remember the many large advertisements for Mail Pouch Tobacco that adorned barns in the 1960s and 1970s), others by local groups, and a few by conversion to other uses (for machinery storage, for example, or even for residences).

Roads, from the simplest dirt or gravel efforts to those that have been paved, are integral parts of many agricultural landscapes. They have released farmers from the need to live on the land, have allowed more efficient distribution of products, and have shaped changing patterns of land use in numerous ways. However simply they may begin, all roads in America now lead to its major cities, easing both the exchange of goods from production areas to final consumers and the access that people in cities have to the surrounding countryside (for better or worse).

Service centers or central places provide goods and services to people who live in the surrounding hinterlands; they also serve as collection and processing centers for the products of those same hinterlands. The size and spacing of central places in agricultural regions is primarily a function of the size and spacing of farms and the rural population density; their geometry has long been studied by geographers. Where farm sizes are large and rural densities low, central places are widely scattered. High rural population densities, in comparison, encourage a network of more closely spaced central places. In turn, rural service centers clearly reflect the kinds of activities that predominate within the region; tall silos in grain-producing regions (usually located along rail lines) and packing sheds and canneries in fruit- and vegetable-producing regions are common examples. Rural service centers also serve as homes for agricultural laborers, some landowners, and a growing number of commuters, in some cases.

Aside from the cultural landscape, geographers have been interested in other aspects of agriculture. Major interests have focused on the spatial distribution of agricultural activities, on areal associations (or correlations) between agricultural activities and other variables that might help explain observed agricultural distributions, and on the ways that these different variables may be interconnected. Describing agricultural regions has also occupied geographers, as has the study of rural populations, from basic population distributions to various types of settlement patterns (dispersed, versus nucleated, for example). Still other geographers have interested themselves in the markets and fairs that are typical of many agricultural areas, in changes that are occurring in those areas (especially as urban refugees move in), and even in the impact on those areas of new patterns of income distribution or leisure.

Still another focus of agricultural geographers has been the study of specialization and its role in the development of significant agricultural regions. As John Fraser Hart (1975) has commented, "Within a fairly extensive area of more-or-less similar conditions the majority of farmers, being rational men, tend to make more-or-less similar decisions, thereby creating an agricultural region." As noted already, such a region, created by specialization, in turn tends to foster a cultural landscape that includes related activities, particular landscape characteristics that set it aside from other regions, and in some cases, even an appeal for those who work within it or visit it.

So far agricultural landscapes have been considered only in a generic sense, by suggesting some of the features that geographers would look for in an effort to better comprehend what makes one agricultural region different from another and what observers of such landscapes might find of interest. We can take many of these ideas now and apply them, first generally and then more specifically, to viticultural landscapes in the United States.

Viticultural Landscapes

Considerable variations can be discerned among viticultural landscapes in the United States. Contributing to these differences are at
least the following: the length of time that viticulture has been practiced in the region, the degree to which viticulture dominates the region, and the natural conditions that allow viticulture to exist. However, before considering differences among them, we should consider
some general characteristics that would help distinguish viticultural
regions and winescapes from other types of agricultural landscapes.
We can do this by considering the following elements: vines and vineyard landscapes, buildings, other artifacts related to viticulture, roads,
central places, and names that express the influence of wine country
on the cultural landscape. Careful observation in viticultural landscapes will increase the rewards of visiting them.

Vines and Vineyard Landscapes

Grapevines everywhere share at least a few common characteristics.
To the untrained (and even sometimes the fairly well-trained) eye,
grapevines tend to look nearly alike, especially before mature grapes
are hanging in heavy bunches from them—drive by Merlot and
Cabernet Sauvignon vines, for example, and you are unlikely to notice the change from one to the other. Grapevines all go dormant in
winter, giving up their leaves, sometimes in a flurry of gold and orange; they all begin to grow again in the spring, when average temperatures again reach approximately 50 degrees Fahrenheit. While
dormant, they all need to be pruned back to selected buds, which will
then produce the next season's crop.

As geographer Harm de Blij (1983:150) once observed, "The vine
clothes the landscape and remakes it." Although the succulent green
of the vines nicely complements the many natural greens of summer
in the humid eastern portions of the nation, this verdancy frequently
stands in stark contrast to the desiccated land and often lifeless summer vegetation surrounding vineyards in the arid, semiarid, and dry
summer Mediterranean climates of the West. Vines adapt easily to
the flat alluvial lands of the Great Central Valley in California; they
follow contours or simply march in single file over the rolling landscapes of low hills from California to New York; and they even sink

roots into the thin soils of steeper slopes, where terraces may be needed to trap runoff water, prevent soil erosion, and provide footing for the workers who tend the vines and harvest the heavy bunches of ripe grapes. Such vines may struggle against all odds, yet they end up producing small crops of intensely flavored grapes that are the delight of dedicated wine makers and consumers alike.

Because they are adaptable, grapevines create landscapes in which the natural contours of the land play a considerable role. Until recently, however, most American vineyards were planted on flat or gently rolling alluvial soils. Now, hillside vineyards are becoming more common and are often even encouraged, as wine makers seek high-quality grapes wherever they can find them.

One of the most spectacular hillside vineyard sites in America is that of Renaissance Vineyards and Winery in California's North Yuba AVA. Growing on terraced hillsides, at elevations of from 1,800 to 2,300 feet, nearly 400 acres of grapevines grow on iron-rich basaltic soils in the western foothills of the Sierra Nevada. Changes in elevation, combined with variations in slope aspect (the direction in which the slope faces) and cold air drainage, create a variety of microclimates within a small area. Skeptics need only consider wine maker Gideon Beinstock's observation (backed by the consistent high quality of his wines) that within very short distances (only a few hundred feet) both Riesling and Cabernet Sauvignon ripen to perfection!

Although at first sight grapevines and topography together seem to play out the same pattern over and over again, such is not always the case; vineyard landscapes can differ considerably for a variety of other reasons. To begin with, the visual impression of a viticultural landscape is greatly affected by the degree to which grapevines share the landscape with other plants, either native or domestic, as well as by the nature of the local climate. Whereas grapevines are almost all you can see anymore on the floor of the Napa Valley (with the exception of a bit of riparian forest along the Napa River and a few small patches of olives and other trees), grapevines in the Yakima Valley share neighborhoods with apples and cherries, and vineyards in small arid valleys in Arizona and New Mexico are bounded more by scattered cacti and other xerophytes. Local landscape conditions are often a source of considerable pride: Joan Wolverton, for example, once pointed out that the vineyard and winery she and her husband farm and own (Salishan Vineyards, Washington) was probably the closest vineyard in the United States to an active volcano. Salishan is about thirty miles from Mount Saint Helens, which erupted in 1980 and which also, by the way, should not be confused with California's Mount St. Helena, that grand volcanic sentinel at the northwest end of the Napa Valley.

Human decisions about viticulture may also considerably affect the look of vineyard landscapes. Two major decisions are the density with which vines are planted and the way in which they are subsequently trained and pruned. At one time, most California vineyards, for example, were planted at a density of about 450 vines per acre (12 feet between rows and 8 feet between vines within the rows). Recently in the Napa Valley, Opus One has been planting Cabernet Sauvignon at 2,200 vines per acre, and some wineries are considering densities of as high as 3,000 vines per acre (French vineyards are often planted at densities of 2,000 or more vines per acre, though overall yields remain closely controlled). Needless to say, the visual appearance of vineyards changes considerably as densities increase.

Aside from choices about grape varieties and vine spacing, the different appearance of vineyards from one place to another is also greatly affected by different growing practices, such as how the vines are trained and pruned. Although there are numerous different ways to train and prune grapevines, the most basic subdivision among the different methods is that of head-training and cordon-training; most systems are variations on one or the other of these two classic ways of training vines.

There are manifold differences between the training and pruning systems used in European wine regions and those most commonly used in American vineyards, but in each case "canopy management" is critical to the resultant quality of the grapes and the wines produced from them. In many German wine regions, for example, trellises are rare. Often growing on steep slopes above the Moselle and Rhine Rivers, grapevines are treated more like small trees, each standing alone and generating its own canopy of leaves (similar individual grapevines can still be seen occasionally in American vineyards as well, though they have become a small minority as trellis systems have become more common). In contrast, the grapevines of Bordeaux, like most of those in America, tend to be on trellises, though in Bordeaux the vines and their canopies tend to be kept lower to the ground (they are also planted at much higher densities per acre than in most American vineyards, though that may be changing in this country, as was noted already). The leaf canopy along a trellis system is often referred to as a curtain, and there are both single- and double-curtain systems in use in different American vineyards. In any case, the leaf canopy is designed (or managed) to control the amount of sunlight reaching the layers of leaves and the grapes, which determines the resultant quality of those grapes.

Leaves are necessary for photosynthesis, of course, but vines that are going to produce excellent wine grapes cannot afford to put all of their energy into producing leaves—too many leaves are not a good thing, though too few threaten the vine's ability to thrive. Viticulturists seek an optimum arrangement of leaves and must take into consideration both plant health (sufficient photosynthesis, among other things) and grape quality. For any given cultivar, some simple relationships between leaf canopy, sunlight, and grape characteristics can be described. For example, the amount of sunlight (or conversely, the degree of shade) will affect sugar levels, levels of malic and tartaric acids and resultant pHs of the grapes, and the degree or depth of coloring in red grape varieties. In turn, the amount of sunlight reaching an individual vine is affected not only by its own leaf canopy but also by that of neighboring vines. This means that the height of trellis systems above ground level and the distance between rows are further variables that must be carefully considered by the viticulturist who is seeking to maximize the quality of wine grapes in the vineyards.

Continued experimentation suggests that there is not yet an optimum system that works everywhere, nor is there one that works equally well with different cultivars. Some varieties, Sauvignon Blanc, for example, can have notably different characteristics within the same growing region depending upon how viticulturists manage the leaf canopy. As you drive through wine country anywhere in the United States, look closely at how different growers and different regions use different trellising, training, and pruning methods to shape the vines and the concomitant winescape.

Aside from the decision about whether to grow vines on trellises, the other major decision that must be made in the vineyard is how to train and prune the vines. When vines are grown without trellises (Figure 6.1), they are virtually always head pruned—that is, they are pruned back each year to a few select spurs or canes, and those spurs or canes contain the buds that will produce the next year's crop of grapes. Spurs are shorter than canes, so they contain fewer buds;

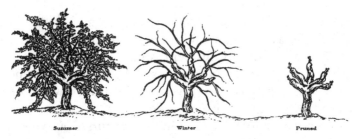

Summer   Winter   Pruned

FIGURE 6.1   Head-Pruned Vines
These vines are typical of earlier plantings in California and can still be seen in a few places in Napa, Sonoma, and San Joaquin Counties. Courtesy of Benziger Family Winery Vineyard Discovery Center, Glen Ellen, California.

spurs usually have between two and five buds, whereas canes are more likely to have between ten and twelve. In any case, careful pruning is essential. It allows control of next year's crop and helps adjust the vine's ratio of leaves to berries. Furthermore, if a vine is overcropped one year, it is likely to have reduced fruitfulness in the following year, so it is preferable to prune vines adeptly each year in order to have consistent crops (within the normal weather variations) of high-quality fruit—balance in the vineyards is often the byword.

Grapevines grown on trellises may also be head-pruned, in which case two or four (or less often, some other number) spurs or, more likely, canes are selected for the production of next year's crop. If canes are selected, they can then be tied onto the trellis system after pruning and will form the basis for next year's leaf canopy and grape production. For some cultivars grown in the colder parts of the northeastern United States, multiple trunks are sometimes encouraged. Each year, a new cane is grown from a spur at the base of the trunk, and after four or five years, that cane will become the new "trunk." Such an arrangement also provides backup trunks in case of a severe freeze, which might kill some but not all of the canes (there are typically five canes, all in different stages of development).

An alternative method of production is to use cordon-trained vines, where two (or sometimes four) canes are allowed to develop into much thicker "branches," or cordons, over a period of years. These grow thicker, more like trunks, but are stretched out along trellis wires like woody arms (usually there are either two or four). The cordons then form the basis of the trellis system; two cordons form a bilateral cordon system, and four form a quadrilateral one. Pruning then normally requires selecting spurs or canes along each cordon. If canes are chosen, they will usually be tied to wires along the trellis

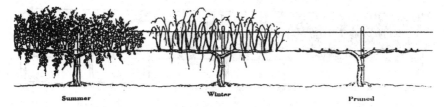

FIGURE 6.2 The Two-Wire Trellis
This trellis system, which allows planting about 450 vines per acre, generally increased production over that of head-pruned vines and was the first important step forward in California trellising. The vine here is shown pruned back to six spurs on each cordon. Courtesy of Benziger Family Winery Vineyard Discovery Center, Glen Ellen, California.

above the cordons, where they will form the basis for next year's crop of leaves and grapes. As Jeff Cox (1985:75) observed, "The best canes for fruit production are those that received good sun exposure in the previous season and are between the size of a pencil and a little finger, depending on the inherent vigor of the vine." Sometimes spurs rather than canes are used, in which case each cordon will be cut back until it has about four to six spurs (Figure 6.2).

In recent years, the simple two-wire trellis system shown in Figure 6.2 has been replaced in many cases with a vertical trellis system, which still uses bilateral cordons but improves canopy management and grape quality (Figure 6.3)—it also gives vineyards a very different visual appearance. In California some variation on the lyre, or "U," trellis (Figure 6.4) is now appearing more frequently, and it employs quadrilateral cordons (effectively almost doubling yields, all else be-

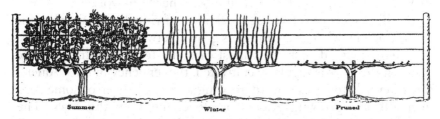

FIGURE 6.3 The Vertical Trellis
This is one of several new systems appearing in California and elsewhere. New shoots are trained upward, allowing sunlight to fall directly on the fruit, produced near the bottom wire. Vines can be planted at much higher densities, often as many as 1,100 per acre. As canopy management becomes more important, this system and newer variations of it are becoming common. Courtesy of Benziger Family Winery Vineyard Discovery Center, Glen Ellen, California.

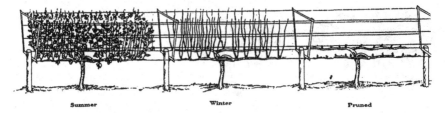

Summer   Winter   Pruned

FIGURE 6.4   The Lyre, or "U," Trellis
With this method, viticulturists create two rows of vines from one by shifting from a bilateral to a quadrilateral cordon system, training the shoots from each pair of cordons upward along vertical wires, much as in the vertical trellis method. Planting densities are normally lower, sometimes only 450 vines per acre. The two extra cordons per vine act to double the output of the vine, and each vertical side of the "U" results in excellent control of sunlight. Courtesy of Benziger Family Winery Vineyard Discovery Center, Glen Ellen, California.

ing equal, since each vine now has four arms rather than two). Many eastern vineyards utilize the Geneva double-curtain (Figure 6.5), which exposes the grapes at the top and also employs quadrilateral cordons. As you can envision, vineyards and winescapes can vary considerably in their general appearance. Other trellis systems include the "T" top, Scott Henry, Te Kauwhata two-tier, and a variety of unique hybrids designed by viticulturists for specific purposes.

Different cultivars and different climatic conditions help determine what type of training and pruning system a grower should employ, and initial pruning and training must be followed by further adjustments in spring, when the vines begin to grow; leaf thinning, for example, may be used to open up the canopy and allow more sunlight to reach the grapes. In any case, pruning requires skilled workers, and they are generally paid better than most other farm workers.

Another aspect of the visual appearance of viticultural landscapes is determined by what, if anything, is grown between the rows of vines. Although vineyards may look most carefully tended when the space between the rows is regularly cultivated, the tendency in most American vineyards today is toward growing a ground cover of some sort, normally low-growing grasses or clover, between the rows. The ground cover serves both to conserve moisture (especially important in areas where summer rains are sparse or nonexistent) and to provide a habitat for desirable insects (especially those that feed on grapevine pests). Often described by terms such as "sustainability" or "greening," such adjustments in the vineyards also make good economic

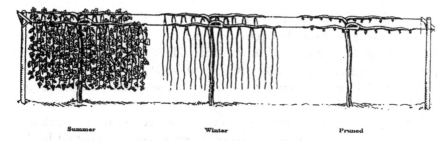

Summer   Winter   Pruned

FIGURE 6.5   The Geneva Double-Curtain Trellis System
This system was developed in New York state and is used more there than in the West. Some viticulturists argue that despite apparent differences, the "U," or lyre, system is really just a variation on the double-curtain system. However, the two systems differ in the position of the grapes, and the lyre system provides better control of sunlight. Courtesy of Benziger Family Winery Vineyard Discovery Center, Glen Ellen, California.

sense—using less water and smaller applications of expensive chemicals is good for everyone involved, including the final consumer.

Look within viticultural landscapes for other evidence of growing practices as well, from the type of irrigation method used (overhead sprinklers or drip systems are most common) to various means of frost control (including overhead sprinklers, wind machines, and smudge pots) and even the direction in which vine rows run. Vine alignments often differ from one region to the next. At higher latitudes (in states such as Washington and New York), it becomes more important to align vines in order to maximize the amount of sunlight caught by the leaf canopy, whereas in California, vine rows tend to run in all directions. Even there, however, some evidence suggests that certain alignments may be better than others. For example, at Opus One in the Napa Valley newer plantings are aligned to maximize air flow through the vineyards, hastening the drying of wet vines and soils during the growing season.

A more melancholy sight is the occasional vineyard that has been left untended, usually adjacent to expanding urban areas. Around Fontana in southern California, for example, such neglected vineyards have been common scenes during the last three decades as the Los Angeles Basin has filled with residents, many of whom seek lower-cost homes ever further from the city's center. Abandoned wineries can still occasionally be seen as well, though most have given way to the restless bulldozers, those modern harbingers of progress.

At the other extreme, however, are pioneering viticultural efforts such as those made by Emil Tedeschi, who planted vines on the slopes of Maui's Haleakala in the 1970s, or Brian Helsaple, who is currently planting vines in California's tiny Seiad Valley, which lies west of Yreka in Siskiyou County and represents the northern frontier of California viticulture.

Finally, and perhaps most obviously, the degree to which viticulture dominates within a region is a major determinant of what that landscape is going to look like. In California, the Napa, Dry Creek, and Alexander Valleys are places where viticulture rules the landscape. In comparison, viticultural landscapes in Texas or Virginia are shaped more both by other crops and by natural vegetation. Whatever the degree of domination, however, viticulture adds its own distinctive touch to landscapes.

Types of Buildings

Buildings in viticultural landscapes can most simply be categorized into two groups, wineries and everything else. Wineries are distinctive features of such landscapes and are also tourist attractions in many regions. Architecturally, they range from simple, functional buildings designed by local contractors to magnificent buildings reminiscent of the grandeur of great French *châteaux*. Older buildings were also often made of native building materials, such as local stone.

Examples of the more elaborate buildings are perhaps most frequently encountered in the Napa Valley, ranging from the great Rheinhaus at Beringer to an assortment of splendid architectural gems such as Sterling (on its handsome hillside site overlooking the vast green carpet of vines that stretches away from it in all directions), the mission-style winery of Robert Mondavi (with its beckoning arch), the soft pastels and unusual shape of Clos Pegase, and the sheer opulence of Opus One—designed to celebrate the coming together of French (Mouton) and American (Mondavi) influences. Elsewhere, American wineries occasionally reach magnificent proportions and architectural distinction as well, as with Chateau St. Jean, Domaine Carneros, and Ferrari-Carano in Sonoma County (California); Domaine Drouhin (Oregon); Chateau Ste. Michelle in Woodinville (Washington); and Prince Michel (Virginia).

Most American wineries, however, fall somewhere below magnificent on an architectural scale. They are functional buildings, typically

The Robert Mondavi Winery in California's Napa Valley.

including not only wine-making facilities but also an inviting tasting room for visitors. Although not necessarily opulent, many are extremely attractive and inviting: Matanzas Creek Winery in Sonoma County (with its wonderful field of lavender in front), Navarro in California's Anderson Valley (which always has a spectacular display of flowers in its garden), and Idaho's Ste. Chapelle come to mind immediately. Wine growing remains a business, of course, and economic constraints to a large extent dictate functionality over extravagance when it comes to building wineries and related structures. Nonetheless, most winery owners are aware that visitors are an important source of potential revenue; well-kept grounds (often wonderfully landscaped), and cordial tasting rooms with knowledgeable personnel are more important to many visitors than ostentatious displays and expensive wines.

Different still are wineries that are currently housed in buildings that were once designed for other purposes—viticultural landscapes often contain vestiges of previous land uses. Dairy barns, for example, have often been converted to wineries: Carey Cellars in California's Santa Ynez Valley is an example. Marty Griffin's Hop Kiln Winery in Sonoma County makes use of an old hop kiln (and a designated state historical landmark), a reminder of an earlier era when hops were a major industry in parts of the county; farther north, in Mendocino County, Milano is also located within an old hop kiln. Even more unusual, Ballard Canyon Winery in California's Santa Ynez Valley moved an abandoned service station onto its vineyard site to serve as the basis for a winery; in the same vein, Milliaire Winery in Murphys, California, is located in an old service station near downtown.

Travel through wine country offers an endless array of interesting lessons to those who look closely, linger a while to ponder the landscape, and ask questions as they go. Some old winery buildings have also been recycled: two Napa Valley examples include the old Christian Brothers Greystone Cellars (built of native stone in 1888), now the western home of the Culinary Institute of America, and the venerable Groezinger Winery (built in 1870), now a tourist shopping complex in Yountville called Vintage 1870. Both are worth visits, and the Culinary Institute runs a fine restaurant that serves lunch and dinner; the menu is Mediterranean, and an open kitchen allows visitors to watch food being prepared and even to talk with the chefs.

Finally, the scale of wine growing shapes a region's buildings and viticultural landscapes as well. For example, in California's Great Central Valley (America's equivalent of Provence and the south of France with respect to wine growing), there are vast vineyards to be seen, but few wineries. E. and J. Gallo, the world's largest winery, is housed in suburban Modesto (not nestled within a sea of vineyards) and looks more like a petroleum refinery than a winery. Visitors are not welcome here, nor are there even signs to indicate either the winery's function or its ownership. JFJ Bronco, Heublein, Giumarra, and other corporate wine-making facilities are also located in or near Central Valley towns and most of them are large, functional, and closed to visitors.

Other Artifacts Related to Viticulture

Homes are another important type of building in viticultural landscapes, though they are far less reflective of such landscapes than are wineries. More often than not, winery owners do not live on their properties, but many vineyards are owned by individual farmers, not wineries, and those farmers often do live on their property. Homes in these areas range from modest to absolutely magnificent, and they tell us something about the wealth generated within the region. Because vineyard landscapes are desirable places to live, there is increasing pressure to build housing for new residents, most of whom have nothing to do with the wine industry. In turn, such suburbanization of

viticultural landscapes threatens to destroy them if they are not protected as agricultural preserves (as is the case in the Napa Valley, for
example). Too often, viticultural landscapes have developed near
growing metropolitan areas, from which commuter sheds range out
ever farther like tentacles reaching out along access corridors. Suburbanization, of course, drives up land values, making agricultural pursuits such as viticulture next to impossible. Older homes in some wine
regions are being converted to use as bed-and-breakfast inns, appealing to wine country visitors with everything from quiet quarters and
easy access to vineyard landscapes to wine tastings, sumptuous break-
fasts, and even offers to pack picnic lunches.

Equipment sheds are common sights, though hardly distinguish-
able from other types of utilitarian agricultural buildings. Tractors,
discs, spray rigs, and mechanical harvesting equipment are best kept
inside, especially during rainy months.

Where viticulture is the dominant activity in a region, fences are
uncommon (though occasionally tall fences around vineyards are
built to discourage hungry deer), and when present, they are more
likely to be decorative than functional. In places where livestock is
also raised, fences (commonly barbed wire) separate vines from animals. Roses or other plants are often planted at the ends of rows of
vines, mainly for decorative purposes; a few California growers are
now planting prune trees at the ends of rows of vines in order to encourage the presence of beneficial insects.

Also visible in many wine country landscapes is outdoor wine-making equipment, some of it in use, some vestigial and left for decora-
tion. Most common are barrels, fermentation tanks, and winepresses.
These are most often seen at or near wineries.

Wine Roads

John B. Jackson (1994:viii), one of the most eloquent students of
American landscapes, noted recently that "I have come back convinced beyond a doubt that much of our contemporary landscape can
no longer be seen as a composition of well-defined individual spaces
... but as the zones of influence and control of roads, streets, highways: arteries which dominate and nourish and hold a landscape together and provide it with instant accessibility." Although Jackson was
hardly speaking of viticultural landscapes, the statement deserves our
attention.

Perhaps the most important consideration is the role that roads play in the location of wineries (many of these arteries, such as California's Russian River Wine Road, have attained the status of what the French call *routes du vin*). As visitors become more common and numerous in wine country, wineries often find some disadvantage in out-of-the-way locations; a look at the distribution of wineries in the Napa Valley, for example, shows the immediate locational attraction of Highway 29 (and to only a somewhat lesser extent, the Silverado Trail that parallels Highway 29). Off these two primary routes, visitors are fewer, though perhaps more dedicated. Wine roads are becoming more common features in winescapes from one end of the nation to the other. Highway 29 through the Napa Valley is still America's quintessential wine road, but others (often with signs pointing to different wineries) carry visitors through vineyard landscapes with equal enthusiasm.

Central Places

Central places serve both as service centers for surrounding populations (within a local hinterland or trade area) and as collection points for agricultural products produced within the trade area, from which they are distributed to regional, national, and even international markets. Aside from common functions performed by most small service centers—the local provision of gasoline, groceries, clothing, furniture, and a host of other necessary goods and services—wine country central places also contribute some distinctive features.

Because of the increased visitation to wine country locations, tourist facilities have become more common (out of proportion to the size of the community), especially lodging facilities, spas, and eateries. Aside from the bed-and-breakfasts mentioned already, local motels often cater to wine country visitors—some offer a bottle of wine with the room, others offer vouchers good for free tastings and tours (at facilities that have a charge). Because of the association between good wine and fine dining, eating facilities in wine country have been improving considerably in recent years, and continued improvements seem likely. Although the menus at these eateries may not thematically reflect their wine country locations, most feature local wines, many of which may be only locally available or may be in short supply.

Names with Wine Country Themes

Names of places within viticultural landscapes often reflect wine country themes, reminding us that a sense of place can be reinforced in a variety of ways. Names of businesses (even many that are completely unrelated to the wine industry), for example, often reflect their wine country locations. Names that include wine, wine country, and vintage are commonly used, though the generic terms "grape" and "grapevine," along with occasional uses of the names of specific cultivars, can also be found. A few of many examples in the Napa Valley area alone include the following: The Vintage Bank, Vintage Academy of Hair Design, and Wine Country Computing. In nearby Sonoma County, examples include the following: Grape Leaf Inn, Grapevine Express, Vine Street Gas Station, Vineyard Bargain Antiques, Vineyard Electronics, Vintage Auto Repair, Wine Country Auto Service, and Vintners Inn. Even whole shopping centers are sometimes given names reminiscent of winescape themes: Vineyard Plaza in Healdsburg, California, for example, and the Vineyard Mall in Lodi, California (which, ironically, is built on land once covered with grapevines!).

Wine-related names are common elsewhere as well. For example, in tiny North East, Pennsylvania, you can stay at the Grape Arbor Inn or Vineyard Bed and Breakfast, eat at Stahlman's Grapevine Cafe, and visit such places as Grape Expectations, Ye Olde Vineyard—Fine Arts and Crafts, and the Grape City Beer Co. There are also two housing divisions of toponymic interest in North East—Vineyard Park and Vineyard Village.

Road and street names may also express wine country relationships. Grape Street, for example, can be found in places as disparate as Hammondsport, New York, and Fresno, California. Just outside of Fresno there is Muscat Road. In the Napa Valley, a St. Helena westside subdivision includes many streets named after cultivars, including Riesling Way, Columbard Court, Chardonnay Way, Sylvaner Avenue, and Pinot Way; there is also a Chablis Circle. Also in St. Helena there are the Vineyard Valley Mobile Home Park and the Vintage Hall Napa Valley Museum. Over in Sonoma County, Petaluma, even though not a central location for the wine industry, has streets such as Chardonnay Lane, Zinfandel Court, Vintage Way, Champagne Place, and Vineyard Drive. On the other side of the nation, in North East, Pennsylvania, there are streets such as Catawba Drive, Isabella Street, Niagara Street, and Vine Street.

Vineyard Way, Vineyard Drive, Vineyard Avenue, and Vineyard Road are names seen frequently in wine country; examples include Vineyard Way (near El Nido, Merced County, California), Vineyard Canyon Road (between San Miguel and Parkfield, California), Vine Avenue (just north of Atwater, California), Vine Hill Road (near Scotts Valley, California), Vineland Avenue (on the west side of Fresno, California), Vineyard Avenue (at least three in California alone—one in Alameda County, one in San Bernardino County, and one in Ventura County), and there are at least four Vineyard Roads in the Golden State alone (in Sacramento, San Joaquin, Stanislaus, and Yuba Counties). Chico, California, has a Grape Way (though it has very few grapevines and only one winery), Lodi has a Tokay Colony Road, Kern County, California, has a Champagne Avenue (near Petroleum Road and far from anyone who makes champagne), and there is a Chianti Road near Healdsburg in Sonoma County.

Sometimes wine-related names are only historical vestiges of landscapes that have lost their viticultural significance. In Los Angeles, for example, near downtown and Union Station, is Vignes Street, named after early California winegrower Jean Luis Vignes, whose vineyard once flourished nearby. The city also has a Vine Way, a Vine Avenue, and a Vine Street, though connections with historic winescapes may be tenuous. Names on the landscape are one of many significant visual expressions within winescapes across the nation; they aid in giving a regional identity to such places and in establishing connections between these places and people's sense of belonging.

Even personalized automobile license plates sometimes reflect wine country themes or influences: California examples include ZINFDL and ZINWINE (Zinfandel), ETUDE PN (Etude Winery, Pinot Noir), ETUDE CS (Etude Winery, Cabernet Sauvignon), GRAPE DR (which belongs to St. Helena wine consultant, or "grape doctor," Richard Nagaoka), and PNTNOIR (Pinot Noir).

A Comparison of Two Winescapes

A more detailed look at two different viticultural landscapes should help better illustrate various aspects of such landscapes as well as similarities and differences among them. California's Napa Valley and

A prominent sign greets visitors to the Napa Valley.

New York's Finger Lakes region both have long viticultural histories; differences in their winescapes reflect many cultural and environmental variables and range from the choice of grapevines to grow to the types of wines that are made. California's Sonoma County winescapes could have been considered as well, but they are less compact than the Napa Valley; they are more diverse and include several different viticultural landscapes, including those of the Dry Creek, Alexander, and Russian River Valleys.

The Napa Valley, California

By any measure one might choose, California's Napa Valley is America's quintessential winescape—almost European in feel, ethereal in its charm and effect on visitors, and dominated almost exclusively by the presence of vineyards, wineries, and small service centers dedicated to providing goods and services for residents and wine country tourists alike. For many, even among Californians, "wine country" and Napa Valley are synonymous.

The Napa Valley is located north and slightly east of San Francisco, and stretches northwestward from San Pablo Bay in the south to just beyond the small town of Calistoga. Its vineyards (more than 30,000 acres) stretch from near Cuttings Wharf (just south of the city of Napa, in the cool Los Carneros AVA) to just beyond Calistoga, a distance of about 30 miles. Seldom does this lovely valley exceed 3 miles in width, except at its southern end.

Highway 29 (America's premier wine road) traverses the west side of the Napa Valley between Napa and Calistoga, passing through Yountville, Oakville, Rutherford, and St. Helena along the way. Generally parallel to it, on the east side of this slender valley, is the Silverado Trail, and between the two the valley is crisscrossed by a series of smaller lanes, from Oak Knoll Avenue in the south to Tubbs Lane, just north of Calistoga. Among these smaller roads is Zinfandel Lane, one of many examples of viticultural names in the Napa Valley winescape.

Open only on its south end, the Napa Valley is bounded on the west by the Mayacamas Mountains, on the east by the Vaca Mountains, and on the north end by Mount St. Helena (4,343 feet)—volcanic activity has shaped much of the local landscape and has provided much of the material upon which local soils have formed as well. Elevations along the valley floor range from 18 feet above sea level at Napa to 419 feet above sea level just north of Calistoga. Vineyards dominate the landscape of the valley floor and sweep their way up local slopes, gradually yielding to oaks and occasional conifers.

Despite its relatively small geographic extent, the Napa Valley includes segments of three different viticultural mesoclimates: Regions I, II, and III. From its southern extent, nearest the cooling marine influence of San Pablo Bay, to about Oakville, the valley has a Region I climate. From Rutherford to a mile or two north of St. Helena, the valley is primarily Region II, and from there northward it changes to Region III. Distance from San Pablo Bay plays the major role, of course, in this distribution of growing environments. Locally, however, topography and other factors create many variations on the general climatic theme, creating pockets of warmer or cooler than average microclimates.

The general pattern of growing climates in turn affects the geographic distribution of particular cultivars—Chardonnay and Pinot Noir are much more common in the south end, whereas the area from Oakville northward is planted more in Cabernet Sauvignon, Merlot, Zinfandel, and other red cultivars. The two most important cultivars in Napa County (most of which is included in the Napa Valley AVA, which stretches somewhat beyond the boundary of the valley proper), with more than 10,000 acres each, are Cabernet Sauvignon and Chardonnay; Merlot, Pinot Noir, Sauvignon Blanc, and

Zinfandel each occupy more than 2,000 acres. A smattering of other varietal grapes rounds out the viticultural landscape.

Eleven American Viticultural Areas are represented in Napa County. Aside from its inclusion in the vast North Coast AVA, Napa County includes a number of other AVAs, the largest of which is the Napa Valley AVA. Other AVAs that are either partially or totally located within the county are the following: Los Carneros (which overlaps into Sonoma County to the west), Oakville, Rutherford, Stags Leap District, Spring Mountain District, Mount Veeder, Howell Mountain, Atlas Peak, and Wild Horse Valley (which overlaps with Solano County to the east).

As recently as twenty years ago, most vineyards in the Napa Valley were planted with similar spacings (typically 8 feet apart in rows that were 12 feet apart), producing vine densities of around 450 vines per acre. Most were also trained as bilateral cordons, though some untrellised head-pruned vines remained (and still do for that matter). In recent years, however, the countenance of the Napa Valley viticultural landscape has given way more to a patchwork of different vine densities and trellis systems, mainly as old or phylloxera-infected vines have been removed and replaced. At Opus One, for example, new vines are being planted at 2,200 per acre (and a few vineyards are reportedly planting as many as 3,000 vines per acre). Bilateral cordon training systems are often being replaced with quadrilateral systems, and new trellises range from the Geneva double-curtain to various lyre and other systems, creating considerably more visual variety in the vineyard landscape. Fewer grape varieties have been used as replacements, among them especially Chardonnay, Cabernet Sauvignon, and Merlot, leading to ever more cultivar specialization in the valley.

Along Highway 29 and the Silverado Trail, wineries seem nearly ubiquitous—there are so many that roads that do not lead to nearby wineries often post signs that say "Not a Winery." Over the course of one year, millions of visitors come here, looking mostly for wineries, then for places to stay and eat. Traffic can be vexing on summer weekends, prompting periodic complaints from local residents. Winery architecture varies from extravagant—Opus One, Sterling, Clos Pegase, Robert Mondavi—to the plain and functional. Many of the earliest wineries were made from local stone, including the Greystone Cellars north of St. Helena; other stone features of interest include walls, fences, and bridges. Most wineries welcome visitors, and many

Sterling Vineyards ranks among the most distinctive sites in the Napa Valley.

provide picnic facilities; a few even have delis for those who arrive without picnic supplies.

Napa County has just over 120,000 residents, most of whom do not work in the wine industry; more than one-half of them live in the city of Napa, and most of them reside either in or near the Napa Valley. Aside from the city of Napa (68,000), with its many elegant Victorian homes, the major service centers in the Napa Valley are Yountville (3,500), St. Helena (5,700), and Calistoga (4,800); smaller places include Oakville (with its well-known Oakville Grocery, a noted supplier of picnic supplies) and Rutherford.

As you might imagine, Napa and these smaller service centers contain an array of service facilities for the wine industry, along with many hostelries and restaurants for the constant string of tourists. In addition, there is the controversial Napa Valley Wine Train, which carries tourists between Napa and St. Helena and has plans now to increase its weekly number of trips. Although tourists find it a splendid way to see the valley, some residents oppose it because it disturbs traffic and increases the number of visitors to a place that many already consider overcrowded.

Wine-related names show up almost everywhere in this wine-steeped viticultural landscape. Aside from those mentioned earlier (especially the street names in St. Helena and, of course, Zinfandel Lane), there are numerous others. You can play a round or two of golf at the Chardonnay Golf Club (just outside of Napa), eat at places such as Vintner's Court (Napa), maybe have a little dessert at the Vintage Sweet Shoppe (Napa), and stay at places such as the Wine Valley Lodge, Vineyard Country Inn, or Chablis Lodge in Napa, the Wine Way Inn in Calistoga, or the Vintage Inn in Yountville. If you're not sure about where you want to stay during your visit, give Grape Escape Vacation Rentals a call. If you need some help getting around, you can turn to Cabernet Taxi or the Chardonnay Limousine Service—keep an eye open for Wine Country Avenue while you're out and about. Vineyard tours can be arranged with Napa Vintage Tour Co. or Napa Vintage Touring.

If you want to linger in the area a little longer, try the Chardonnay Apartments in Napa; or, if you're elderly, you might consider Vine Village. In St. Helena, there is the Vineland Vista Mobilehome Park and in Napa, the Wine Valley Mobile Estates. Wondering what to do with your pet? How about the Vintage Dog Palace in Napa? Kids still in school? There is always Napa's Vintage High School for the older ones and the Wine Country Day Preschool for the little ones.

For those who want to send down some roots here, help is available from the following: Vintage Properties, Vintage Ranch Properties, and Vintage Mortgage and Investment Co. Need a little cleanup work accomplished? Grapevine Janitorial and Vintage Janitorial Supplies might come in handy. Wine Country Painting and Sandblasting, Wine Country Window and Door, Wine Valley Landscaping, and Vintage Termite Control Co. might all prove useful as well if you're setting up house here.

Gifts can be found at Vintage Collections, Vintage Concepts, The Vintage Jeweler, Vintage Leather, Grape Shirts of Napa, or the Wine Country Florist. An assortment (nonexhaustive, by the way) of other wine-related names within the Napa Valley include the following: Grapes of Bath, Grapevine Lock and Safe, Cabernet Travel of Napa Valley, Grape Graphics, The Grape Yard Shopping Center, Napa Vintage Storage, Vines Gallery, Grapevine Video, Grapevine Tax Service, Vintage Electric, Vintage Pastimes, Vintage Pool Finishing, Vintage Surgical Supplies, Vintage 1870 (a restored winery with boutique gift shops in Yountville), Wine Country Ceramics, the Wine Country Equestrian Center, the Wine Country Financial Group, and Wine Valley Weddings.

Evidence of the Napa Valley's viticultural heritage is ubiquitous, from the patchwork of vines that carpet the winescape and its world-famous wines to the wine-related names that dot the landscape.

Short of going there, a good place for anyone to start getting information is on the World Wide Web, where you can visit the Napa Valley home page.

The Finger Lakes, New York

Located in upstate New York, the Finger Lakes viticultural region is visually appealing throughout the year, with seasonal changes that are much more dramatic than those of the Napa Valley. As geographer John Baxevanis (1992:131) noted, "The Finger Lakes region is an area of legendary beauty, captivating geology, spectacular waterfalls, small historic towns, outstanding glacial remains, and rolling hills." Peppering the entire region, the best vineyards grow on hillsides that overlook the lakes. Geologically speaking, the Finger Lakes owe their origin to glaciation. Long and relatively deep, some of the larger lakes modify the local climate, creating microclimatic niches within which wine grapes are able to ripen; this has been New York's major viticultural area for well over 100 years. William Warner Bostwick, an Episcopal minister at Hammondsport, planted Isabella and Catawba vines in 1829, and by 1860, there were more than 200 acres of vineyards growing around Keuka Lake alone. The half-dozen major Finger Lakes are Canandaigua, Keuka, Seneca, Cayuga (remember Cornell University's song, "Far Above Cayuga's Waters"), Owasco, and Skaneateles.

Two American Viticultural Areas have been established in this region, Finger Lakes and Cayuga Lake. The larger and broader of the two AVAs is the Finger Lakes, which encompasses eleven lakes and all or parts of ten counties, a total of about 4,000 square miles. Within the Finger Lakes region, there are about 15,000 acres of vines and around fifty wineries, including America's second-largest winery corporation, the Canandaigua Wine Company. The Cayuga Lake AVA lies within the broader Finger Lakes AVA and includes about 600 acres of vines and ten wineries. According to wine writer Jeff Morgan (1995:80), the Finger Lakes region is "a land of down-to-earth growers and wine makers, many of them formally trained at the school of hard knocks." Certainly, neither the wineries nor the vineyards have been the center of massive investment similar to that in the Napa Valley or in many other American wine regions in recent decades. For that reason, a decade ago geographer James Newman (1986:316) concluded an article on this region by saying that "there is no doubt that the Finger Lakes wine industry is in a state of crisis. However, total collapse is unlikely in view of the region's history and some recent successes with wine making." Fortunately, the industry is surviving in the region and producing some very good wines in the process, though the scale remains small.

The general climate of this region is humid continental, but locally the larger lakes modify that climate's more extreme elements. Once mostly forested, one-half or more of the region has been cleared over the decades and is used for agriculture, including viticulture. Most of the vineyards are found on gentle slopes around Canandaigua, Keuka, Seneca, and Cayuga Lakes. The rolling landscape allows cold air to drain downward, primarily onto the lakes; in summer, the lakes remain somewhat cooler than the surrounding landscape. One effect, then, is to delay bud break in the spring (when frost could be problematic); the growing season is also stretched out somewhat around the lakes (to as much as 170 days), and autumn frosts are delayed because the lakes remain warmer than the surrounding land at that time of year. Nonetheless, the microclimates that support viticulture here fall within the Region I mesoclimate, sharply constraining the vine choices of viticulturists. Winter snows actually insulate the roots of the dormant vines and provide moisture for spring's renewed growth.

Unlike the Napa Valley, with its nearly solid carpet of vines, the Finger Lakes vines crowd around localized sites near the major lakes, doing best on slopes that face the lakes. Viticulture shares the land with many other crops and with some grazing livestock as well.

The Concord remains New York's (and the Finger Lakes') leading cultivar (a status that was achieved during Prohibition), followed by Catawba, Niagara, Aurore, and Chardonnay (which has been climbing steadily in acreage, helping American Chardonnay acreage to exceed that in France). American and French-American hybrid varieties still dominate, but *Vitis vinifera* is slowly gaining; after Chardonnay, the next most important of these cultivars are Riesling and Pinot Noir. According to James Newman (1986:316), "The zones of lower elevation, gentle slopes, pronounced lake effect, and high lime soils around lakes Seneca and Cayuga are well suited to the new vines, especially vinifera." White cultivars have fared far better so far than reds in the Finger Lakes region.

Until the 1970s, when the market for their grapes began to shrivel rapidly, most growers produced *Vitis labrusca* grapes for Taylor and other large wineries in the region. Subsequently, many had to find other markets for their grapes, learn to make wine themselves, or find another occupation. Many growers rose to the challenge, and today wines in the region are better than they have ever been—wines from Hermann J. Wiemer, Dr. Konstantin Frank (the region's pioneer experimenter with *vinifera* cultivars), Knapp, Lamoreaux Landing, Wagner, and Glenora provide all the proof that one needs to have this region taken seriously by American wine consumers. According to Jeff Morgan (1995:84), "Both the present and future look good for the Finger Lakes."

Located on the site of a Seneca Indian village that was destroyed in 1779, Canandaigua (population 10,700) is perhaps the leading service center for the grape and wine industry in the Finger Lakes region; it is home to the Canandaigua Wine Company (which is exceeded in wine capacity in this country only by E. and J. Gallo). Visitors are welcome at the winery's Sonnenberg Gardens, which includes a wine-tasting room. Smaller service centers in the Finger Lakes region include Naples, Hammondsport, Dundee, Penn Yan, Watkins Glen, Geneva, Lodi, King Ferry, Branchport, Interlaken, and Ithaca (these mostly very European-sounding names reflect the region's settlement by Europeans, though Canandaigua derives from the Seneca Indians' language). Although it is not as saturated with wine-related names as the Napa Valley, the Finger Lakes winescape has at least a few scattered examples. Canandaigua has its Grape Street, and you can follow the Champagne Trail. Other names around the Finger Lakes region include the following: Grapevine Florist, Vine City Supply, Vinehurst, Wine Country Janitorial, and the Wine Country Tourism Association. Those interested in knowing more about viticulture and other aspects of the Finger Lakes region will also find an abundance of material available on the World Wide Web.

7

$\mathscr{S}$easons, Ceremonies, and Wine-Judging Events

$\mathscr{S}$EASONAL CELEBRATIONS AND CEREMONIES have long been associated with agriculture—gods, from Bacchus and Dionysus to the Mother Goddess (idolized more than 8,000 years ago at Çatal Hüyük on the Konya Plain of Anatolia in present-day Turkey), have had considerable connections to both the annual renewal of the seasons and the fertility of the soil. Contrary to T. S. Eliot's (1963:53) lament that "April is the cruelest month," spring has long been celebrated as the time of sowing, autumn that of reaping. Harvest celebrations and rituals, even the harvest moon (the full moon nearest the autumnal equinox), remind us of our ultimate ties to the earth and its fertility, our ultimate dependence on fruitful combinations of soil, water, and weather for our daily bread, and perhaps even our basic animal nature. Such seasonal celebrations deepen our attachments to place and remind us of the steady march of time. As John B. Jackson (1994:160) phrased it, "The special days for plowing, for planting, for harvesting, the days set aside for honoring the local saint, were days when the local sense of place was most vivid."

The Annual Cycle in a Modern American Vineyard

Seasonal rhythms and related celebrations help furnish viticultural regions with a distinctive sense of place.

Spring

In his poem *Atalanta in Calydon,* Algernon Charles Swinburne (Findlay 1982:10) ends a series of couplets about seasonal changes with the following lines: "And in green underwood and cover/Blossom by blossom the spring begins." This annual renewal, so revered by poets and lovers (along with baseball fans and gardeners), sweetens the air with bloom, adds life and color to landscapes once buried in the drab garb of winter, and brings to the table such spring delicacies as fresh asparagus. So, too, is there renewal in the vineyards; as average daily temperatures reach about 50 degrees Fahrenheit, sap stirs in the vines, buds begin to swell, and soon the first green shoots are thrust forth, to be slowly shaped into the branches, leaves, and berries of the new vine year.

Each new shoot usually contains two bunches of tiny flower buds that will set the new grape crop. In California vineyards, "bud break," as this phenomenon is more technically known, typically occurs in March; in cooler growing areas, especially in the eastern United States, it may be delayed into April. Whenever it occurs, it will be followed within a few weeks by the actual, if not visually engaging, process of blooming—tiny, self-pollinating flowers lend their subtle but distinctive fragrance to the air as the new crop is being set.

As you might assume, this can be a fragile time in the vineyards. Frost, hail, or even hard rains can damage or destroy the tiny flowers and tender shoots, resulting in significant crop losses and diminished fruit quality. Among the myriad problems faced by viticulturists, damage of the new flowering shoots is among the most difficult to manage. As noted earlier, the threatened damage from frost can be mitigated to some extent by using wind machines, sprinklers, or smudge pots; hail, however, can be devastating, and its effects impossible to prevent.

As the vines come to life, there are other tasks to be accomplished in the vineyards. Ground covers may need to be established, mulches removed from around vines in cold areas, and there may perhaps be some plowing or disking to do as well. Once the vines are through flowering, a careful walk through the vineyards may suggest the need for some further thinning, more tying of new shoots to trellis wires, and spraying (though in American vineyards viticulturists are trying to wean themselves from heavy applications of pesticides and herbicides).

Summer

Summer rains are seldom critical in American vineyards because water can be provided to thirsty vines by irrigation, if need be, as is done throughout most western vineyards. More critical to the perfect ripening of the grapes is sunshine—neither too little, nor too much. Remember that as the grapes ripen, acid levels gradually drop, so wherever wine grapes are grown they need enough sun and warmth to gradually ripen but not so much that they lose their delicate balance between sugar and acid levels.

Green forests of grapevines create landscapes of considerable charm, especially where summer rainfall is lacking and the natural landscape appears parched and thirsty, surrounding the verdant rows of vines like a weathered picture frame. Throughout much of the summer, there are tasks to be performed in the vineyards—weed control, spraying, and insect management are examples. Leaf canopies need to be managed as well, to make sure that leaves are well positioned and that grapes get adequate amounts of sunlight. As summer progresses, *véraison* occurs; at this important point in their development, grapes increase rapidly in size (gaining sugar and water, which in turn dilutes their acidity), begin to soften, and take on a new color. At *véraison*, white cultivars turn more yellow or gold; red cultivars, of course, turn red (and sometimes dark purple or even almost black). The pace in the vineyard quickens a bit as the harvest approaches. In the wineries, preparations must be made as well, cleaning vats and barrels, making sure crushers and presses are operating, and watching grape sugar levels carefully in order to determine when the harvest should begin (different varieties ripen at different times, of course).

Fall

From late summer on into fall, the grapes are ready to harvest. Once it has sufficiently ripened (with appropriate sugar levels and acidity), the fruit can no longer wait. As a result, harvest time can be frantic with activity; the demand for pickers rises dramatically and fields fill with the sight and sound of people working, with the mingled voices of different languages being not at all unusual. Tubs of ripe grapes are emptied into gondolas, which then wend their way from the vineyards to the waiting crushers at local wineries. In the wineries, workers move to transfer loads of grapes to crushers, watch for indications of ripeness (as well as mold or other problems), and prepare to begin the magical transformation of grapes into wine. There is little rest during harvest time—in some areas night harvesting is carried on in order to avoid the damaging heat of the day, especially when machines can be used to strip grapes from the vines.

Because different cultivars ripen at different times, the harvest can linger in a region for several weeks. Early-ripening varieties are already bubbling away in fermentation tanks as later-ripening varieties wait to be picked. Within the winery, arrays of hoses, collectively referred to sometimes as "spaghetti," wend their way along floors sticky with spilled juice, connecting fermentation tanks to other tanks or barrels, within which the new wine will begin to mature.

Winter

In America's viticultural landscapes, winter cold may be problematic in places such as Hammondsport and Canandaigua in New York but is much less of a problem, if it is at all, in Napa, Paso Robles, or Temecula in California. Winter vineyard landscapes do not all look alike, even though the grapevines are dormant within them. Whereas February could easily bring vines standing above a soft covering of snow in New York's Finger Lakes region, Napa vineyards are more likely to be riotous with the wonderful yellow of wild mustard in full bloom amid the somber trunks of dormant vines. The Napa Valley in late winter is home to the annual Mustard Festival, a harbinger of spring and a good excuse to rouse visitors from their winter lull in the nearby Bay Area. Crafts, foods, wines, and art combine in places such as Yountville's Vintage 1870—residents and visitors to wine country need little excuse for a celebration or festival.

Despite winter's interruption of the vine's life cycle, dormant vines still require attention from growers. Most important, winter is the time for pruning, for carefully selecting the healthy spurs or canes that will produce next year's crop. Pruning is a skilled craft, and as mentioned, pruners are better paid than most agricultural workers. We might envision a scene that is repeated throughout the Napa Valley in winter. A small crew of workers is busily pruning Zinfandel vines: Pruners speak quietly and sparingly as they work their way down rows of winter vines that have canes going in all directions. Each worker takes a row, studies each vine before cutting away all but the strongest canes, then moves on. Behind them, another group follows, tying the selected canes onto the trellis wires. Order must be restored to the vineyards in preparation for the spring to come.

Each season among the vines has its own character, renders a slightly different appearance to the landscape, suggests a different rhythm of work—and each season offers visitors and residents alike a different perspective. Within America's many winescapes, the words of poet William Browne (Hazlitt 1970) ring true: "There is no season such delight can bring, /As summer, autumn, winter, and the spring."

Local Festivals, Ceremonies, and Celebrations

Although people, until recently primarily agrarian, once celebrated planting and harvesting as a central feature of their lives and livelihood, today's festivals, ceremonies, and celebrations in American wine country are less tied to either the calendar or livelihood than they once were. They have also multiplied considerably in number in recent decades, and have expanded from well-known vineyard regions such as the Napa Valley and the Finger Lakes to all but the most remote viticultural sites. Food and wine remain central parts of most events, along with music; more recent additions include hot-air balloons and grape stompings. In any case, most of these celebrations in American wine country today seem designed more to attract tourists, who bolster the local economy by buying wine and food, staying in lodgings, and purchasing souvenirs. Local communities promote these events to a considerable degree, as some examples will make apparent; wineries also sponsor their own events, from concerts and picnics in the vineyards to formal wine tastings and auctions. Although these events have become too numerous to treat individually here, a sampling from America's wine regions will provide a taste of what you can expect to find if you sojourn into wine country. These events, of course, help deepen the sense of place in winescapes and add tradition and meaning to the lives of those who live within them.

California

As you might presume, California's diverse wine regions host numerous celebrations that are related to wine. In southern California, Temecula offers the Temecula Valley Balloon and Wine Festival in May and features a Nouveau Celebration in November.

Within the South Central Coast wine region (Santa Barbara and San Luis Obispo Counties) are such events as the Santa Barbara County Vintners' Festival in April, the Paso Robles Wine Festival in May, the Ojai Wine Festival in June, the Central Coast Wine Festival (in San Luis Obispo) in September, and A Celebration of Harvest (in Santa Ynez) in October.

The North Central Coast wine region (from Monterey County to the San Francisco Bay Area) holds such events as the Santa Clara Winegrowers Spring Wine Festival (in Gilroy, home of the famous Garlic Festival as well) in April, the Bluegrass Arts and Wine Festival (in Felton) in May, the Vintners' Festival in June (located at various wineries in the Santa Cruz Mountains AVA), and the Santa Clara Art and Wine Festival in September.

The North Coast wine region (Napa, Sonoma, Mendocino, and Lake Counties) provides a number of interesting events for tourists and residents alike. Winter Wineland in January is celebrated at a number of different Sonoma County wineries, followed in February by the Mustard Festival in the Napa Valley, in March by the Heart of the Valley Barrel Tasting (at several locations within the Sonoma Valley) and the Russian River Wine Road Barrel Tasting (also at various locations), in May by the Russian River Wine Festival and the Summer Chardonnay Celebration. In July, the tiny town of Philo hosts the California Wine Tasting Championships, followed in August by the Sonoma County Showcase and Wine Auction. Just outside of wine country, Winesong is an annual benefit celebration held in Fort Bragg in September—numerous North Coast wineries participate; another September event is Sonoma's Valley of the Moon Vintage Festival. October in the North Coast wine region brings three harvest celebrations: the Harvest Festival and Warehouse Sale (in Kelseyville), the Sonoma County Harvest Fair (in Santa Rosa), and the Crush Festival (at Chateau de Baun in Fulton). November has the Zinful Celebration (at Fritz Cellars in Cloverdale).

California's Great Central Valley and the adjacent foothills of the Sierra Nevada provide a few annual events of interest as well. Among them are the El Dorado County Passport Weekend in March or April, the Lodi Spring Wine Show and Food Faire in April, the Sierra Showcase of Wine (at the Amador County Fairgrounds) in May, the Fairplay Wine Festival in June, and the Delicato Grape Stomp in September. Seasonal festivities end with October's Wine Appreciation Week, which is held in various Amador County wineries. Although not exactly wine related, spring can be a great time to travel along the Fresno County Blossom Trail to make the self-guided sixty-two-mile journey through blooming orchards—almond, apple, plum, citrus, apricot, and peach; spring is blooming time for California poppies, lupine, baby blue eyes, popcorn flowers, and a host of other wildflowers as well.

The Pacific Northwest

During the past three decades, this wine region—including Oregon, Washington, and Idaho—has risen to national prominence. As in California, there are a number of wine country events here that beckon to tourists and locals to participate.

In Washington in April, there is the Yakima Valley Wine Growers Association Spring Barrel Tasting, in which most or all of the valley's wineries participate. Most wineries also join in celebrating Thanksgiving in the Wine Country. In addition, there is the Pacific Northwest Wine Festival in August and the Tri-Cities Northwest Wine Festival (at Pasco) in November. Aside from these four major events, many wineries sponsor their own celebrations as well: Examples include Cherry Blossom Time at Stewart Vineyards, the Annual Wine and Garlic Celebration at Eaton Hill Winery, the summer Country B.B.Q. at Covey Run, the Basil Festival and Pesto Extravaganza at Hinzerling Winery, and harvest festivals in the fall at Oakwood Cellars, Covey Run, Hyatt Vineyards, and Washington Hills. An event that is not exactly wine related is The Great Prosser Balloon Rally and Harvest Festival in September, which does have a few wineries among its sponsors; for the less adventuresome or for those looking for something to do in the dead of winter, there is Yakima's Celebration of Chocolate and Red Wine in February.

Oregon's wineries have found much to celebrate as well. Some examples include February's Valentine Dinner with the Winemaker, the Bacchus Wine Festival, the Memorial Day Winery Tour (in Polk County), and the Memorial Day Barrel Tasting (at various wineries) in May, the Fourth of July Barrel Tasting at area wineries, the Vintage Festival (Yamhill County) in September, and the Harvest Faire in October. In November, there is the Washington County Wineries Holiday Open House as well, along with the annual Wine Country Thanksgiving in Yamhill County. As elsewhere, many wineries hold their own annual events, from musical performances to picnics and dinners.

New York

In the relatively new winescape of eastern Long Island, one annual event has already become well established, the Long Island Barrel Tasting and Barbecue, held in August. Around the Finger Lakes, wine-related events include the following: the Wine and Flower Festival (Keuka Lake) in April, the Wine and Herb Festival in May, Pasta and Wine in June, the Finger Lakes Wine Fest in August, the Keuka Lake Harvest Festival and the Great Naples Grape and Music Festival in September (as well as the Cheese Festival, which might interest wine lovers), the Wine Glass Marathon (Taylor Winery) in October, and in November, the Champagne and Dessert Wine Festival. The Christmas Cheer and Wine Library Auction rounds out the calendar for the Finger Lakes region; by then the vines are dormant, and most often, snow welcomes participants. As elsewhere, many individual wineries in the region (there are close to fifty of them) hold open houses and other celebrations as well.

Virginia

Thomas Jefferson would certainly be proud to see the viticultural progress that this state has made in the last two decades—more than forty wineries are located in Virginia now, making some very good wines. Among many local and regional wine events in Virginia are the following: Spring Barrel Tasting and Blending (Ingleside) and the Food and Wine Spectacular in March, the Caroline County Wine Festival and the Roanoke Valley Wine Festival in April, the Vintage Virginia—Virginia Wineries Festival (well into its second decade and getting better every year) in June, the Eastern Wine Festival in August, the Smith Mountain Lake Wine Festival in September, and the Alexandria Wine and Arts Festival. In addition, individual wineries are involved in an array of events that keep things hopping for visitors: Barrel tastings and harvest festivals abound, sometimes combined (for example, the Barboursville Vineyards Autumn Explosion and Barrel Tasting). Winter even brings a number of soup and wine events—a great idea, especially when accompanied by welcoming hosts or hostesses and a warm fire; this sounds like a great excuse for a change of pace for people living in places such as nearby Washington, D.C.

A Smattering of Other Examples

For starters in this section, how about the Texas New Vintage Wine and Food Festival, held in April in Grapevine, Texas—how much more viticulturally appropriate could something be? In Arizona's

Sonoita wine region, there is the annual Southern Arizona Food, Wine and Music Exposition, benefiting the Heart-Lung Association and sponsored by a combination of local wineries and restaurants. Up in Ohio, Arnie Esterer (Markko Vineyard) holds an annual Blessing of the Vines in June, and there is a fall Grape Festival in Ashtabula County at Geneva. Down in Arkansas, historic Wiederkehr Village hosts each fall the Wiederkehr Winefest.

Even though Americans consume little wine compared to their European counterparts, they certainly enjoy excursions into wine country, wherever they find it. In turn, wineries welcome them, festivals and celebrations lure them, and small towns are finding new ways to survive by encouraging them to visit.

Looking through this sampling of events, you have probably discerned the often-repeated connection between food and wine, and between wine and the arts as well. Because most wine regions also produce a variety of fruits and vegetables, you are likely to find local festivals celebrating that produce also—garlic, asparagus, apples, and peaches all have festive events associated with them, from New York and New Jersey to California. As you visit wine country, be ready to absorb and enjoy a variety of other agricultural landscapes as well, from fragrant blossoms in spring to wonderfully ripe fruits and vegetables in summer and fall. This abundance of fine produce, coupled with regional wines, tempts all but the most mundane of chefs to develop regional specialties, adding still another dimension to our experience of place associated with viticultural landscapes.

Fairs and Other Wine-Judging Events

For some, it is not enough to enjoy the landscapes produced by viticultural pursuits, to casually sip wines among the vines, savor local foods, and lose track of time in relaxed conversation. There is also the competitive urge (not purely American, but persistently strong here), the need to judge one wine against another, the need to know what is best (according to standards that may easily elude the novice and sometimes send even more experienced tasters in search of new adjectives as they spin their favorite flavor wheels).

In areas where wine production has a lengthy history, local and state fair wine judgings have become traditions, and some of them remain among the best in the nation. Aside from fairs, other organized wine judgings are held annually as well. Direct comparability among wine-judging events is hampered by such things as what governs whose wines are entered in the event and the way that judges are chosen. However, much can be learned about wine quality and, by association, about wine regions by looking at the results of major wine-judging events. Each year provides us with a new vintage, a new round of judgings, and new opportunities to evaluate not only the current vintage from one place to another but also the consistency (or inconsistency) of producers and regions.

Among America's premier wine-judging events (those with a national reputation) are the following: California State Fair, Los Angeles County Fair, Orange County Fair, Farmers Fair (Riverside), San Francisco Fair, Dallas Morning News, New World International (in Diamond Bar, California), Reno–West Coast Competition, and San Diego Competition.

Rules governing the entry of wines into judgings mean that some of the competitions just listed are going to be different from others. The following list provides a general idea of the range of wines that are judged in some important venues. The California State Fair is open to any California wine, for example, whereas the Orange County Fair is open only to California wines that are available in Orange County markets (though that includes a majority of commercial wines made in the Golden State). The Reno–West Coast event is open to any wines from California, Oregon, Washington, and Idaho. Any American wine produced in the United States is welcome at the Farmers Fair (Riverside) and at the Dallas Morning News event, whereas any American *vinifera* wine can be entered in the San Diego Competition. As its name suggests, the New World International event is open to any wine produced in the New World. Wines from anywhere in the Americas are welcomed at the Los Angeles County Fair, and those from anywhere in the world are allowed to be entered in the San Francisco Fair wine judging.

Panels of judges for these competitions are chosen in different ways. The Orange County Fair Commercial Wine Competition (as it is officially known, to distinguish it from the fair's home wine-making competition) uses wine professionals—primarily wine makers and winery owners—to judge wine entries. By contrast, the Los Angeles County Fair uses a more varied panel of experts, including wine writers, sommeliers, wine retailers, wine buyers, and wine lovers (ranging from attorneys and physicians to even an occasional professor).

Whatever the case, however, serious wine judgings are careful to put together panels of people who can evaluate wines; this does not mean, however, that any two panels of experts will provide exactly the same ratings for a sample of wines. Not only can the experts differ (because of different palates and different preferences, for example) but the samples themselves can also sometimes differ (consider the same wine tasted in January by one panel and in July by another, for instance).

Results of these and other wine competitions are made available, usually in booklet form, from the event organizers; many of them are also available in wineshops or from Jerry Mead (editor and publisher of the *Wine Trader*, Carson City, Nevada). California wine consumers can find an annual summary of how California wines fared in all of these events in an annual publication from Varietal Fair entitled *California Wine Winners*, which occasionally is available at wine country bookstores or from Jerry Mead.

Although they have become too numerous to include here in any coherent way, both wine auctions and wine and food events have become increasingly popular. Wine auctions are typically either commercial (held at Christie's and Sotheby's, for example) or are held as benefits (raising money for charities, public television, art museums, and many other causes); the magazine *Wine Spectator* publishes an "Auction Calendar" in each issue. Major wine and food events are held annually in places from Aspen and Yosemite to San Francisco and New York City, and "dinners with the wine maker" have become a popular way for restaurants and wineries to market their products.

In addition to annual organized judgings, auctions, and other wine events, many consumer-oriented publications provide wine-tasting results from their own panels of wine tasters, and wine writers render opinions about wines in columns that appear weekly in newspapers across the nation. Whether all of these results of wine tasting and judging are of value to the average American wine consumer is debatable, though wine sales and promotions are often influenced by them. More will be said about these sources in Chapter 9; in the meantime, consider what it might mean to live in or to visit wine country by viewing winescapes as working landscapes.

8

The Viticultural Area as a Working Landscape

⟨T⟩HE LINGERING CORDIAL FEELINGS that persist after a wonderful weekend spent away from the daily grind in some harried metropolis on a visit to the more serene setting of a nearby wine country landscape often obscure a visitor's perception of the working nature of such landscapes. They are, according to journalist Tony Hiss (1990:114–115), "landscapes whose function and look, or character, or feel, have been shaped over time by sequential, ongoing human activities as much as by natural processes." Following tradition, we can describe viticultural (and, for that matter, other agricultural) landscapes as "working landscapes." Perceptions of such landscapes are influenced by our experiences with them; tourists are never going to see winescapes in quite the same light as do those who live and work in them. An in-depth appreciation of any landscape comes only with time, with encounters throughout the sweep of seasonal changes and rhythms, with careful sorting of symbols that give a landscape its sense of place, its unique regional identity.

Working landscapes are sculpted first by nature, then by the persistent labor and decisions of those who own and work the land. These landscapes have an additional value as well, though it is more difficult to measure, more slippery to fit on any economist's scale of cost-benefit analysis, but it may become ever more important in a badgered, overcrowded world. That benefit is a kind of "refuge value," which can be measured (at least in part) by the willingness of visitors to spend money in local restaurants and motels, among other things. Harder to capture, especially in measurable monetary terms, are the benefits that accrue to visitors, aside from the pleasure derived from excellent food, fine wine, and the ephemeral experience of a more leisurely pace. Fresh air, dramatic earthly beauty, stillness, and the measured annual rhythms of many rural landscapes have restorative value for the soul, however elusive attempts to measure such value might be. Increasingly, the chambers of commerce in central places in or near significant viticultural areas see direct economic benefits in keeping such areas attractive to outsiders. The "greening" of wine growing—the gradual transition to sustainable agriculture—promises to be an additional attraction as viticulturists seek more environmentally friendly and endurable methods of producing their luscious crops without hurting either yields or quality.

Living and Working Among the Vines

Unlike visitors, residents of winescapes—land and winery owners, local business owners, agricultural laborers, and a host of others, most of whom do not work in the wine industry—live, work, and finally die in these working viticultural landscapes. For them, the romantic images that visitors bring with them often contrast sharply with their own perceptions. Seldom do visitors to a region comprehend its internal workings or deeply experience its true sense of place; rather, they take from it a small selection of experiences, a biased sample of regional character, shaped mainly by the moment.

For example, weekend visitors to Fess Parker's resplendent winery in California's Santa Ynez Valley are likely to find Fess himself there to greet them (tall, distinguished, and always pleasant, Fess is still known to older visitors as the actor who played Davy Crockett!). At the same time, they are unlikely to notice anyone working in the vineyards or, for that matter, even in the winery (except behind the tasting bar, of course). Most visitors remain unconcerned about the working landscape around them; they are focused mainly on the wine in their tasting glass. But with a little additional effort, they could look around the little wine country towns and quickly see that most who toil among the grapevines are, like other agricultural workers, not well rewarded for their efforts (despite the fact that workers in wine-grape vineyards are better paid than most agricultural laborers).

The Residents of Wine Country

Visitors who are in California wine country (or in many other wine regions in the nation as well) during the week and are willing to get up early enough and drive around local towns will see day laborers gathering in public places, hoping to be picked up for a day's work in the vineyards. They gather early, usually quietly, typically conversing in Spanish. In Healdsburg in Sonoma County, they gather in the town's plaza, often cool in the early morning fog—they work extremely hard, retreat at night to modest and overcrowded dwellings, endure low wages and few benefits, yet seldom openly complain. Like the solemn rows of vines and scattering of wineries, they, too, shape the viticultural landscape, but in ways that most visitors never observe. Although attempts to organize farm labor have been carried on for decades, including some grape boycotts that received national attention, successes have been few and the gains for farm laborers nearly nonexistent. It is a paradox of life in modern America that those who produce most of the nation's agricultural abundance—not just the farm workers but most of the smaller independent farmers as well—are so inadequately rewarded for their backbreaking efforts.

Only if you drive the back streets of the local towns or stroll within the vineyards during the week will you get any impression of the farm laborer's impact on the winescape. Otherwise, their unsung contributions remain only as perceived nuances in a glass of Cabernet Sauvignon or Chardonnay, for which they receive no credit. In contrast, the wine maker (though seldom those who assist in the process) may receive both considerable notoriety and substantial monetary rewards.

Viticulturists are farmers. This may seem obvious by now, but visitors too often fail to appreciate that the wineries seldom own or farm all of the vineyards from which they receive grapes. In many viticultural areas, especially those that have developed more recently, local farmers have frequently shifted from other crops to grapes mainly because grapes were more financially rewarding to them, not out of some inherent love of either grapes or wine. For example, many of California's Sonoma County viticulturists once raised prunes or apples; most are no more emotionally attached to grapevines than they were to their prune orchards. To many, farming is farming, a good way to live (and, if all goes well, to make a living). Their concern is not with a glass of wine but with the price of grapes (which has been rising rapidly in recent years in most American wine-growing areas)—for them, and for virtually everyone else, wine is a business. Romanticism is displaced by the essential "bean counting," by talk of profits and losses. Like farmers everywhere, viticulturists worry about the weather, the size and quality of the crop, and the vagaries of the marketplace.

Wine country also houses wealthy residents. Large landowners, some winery owners, and a few others find life among the vines sufficiently inviting to want to live there. Growing numbers of wealthy people have moved into the Napa Valley, for example, not only because of its enticing and beautiful winescapes but also because of its proximity to the cultural life of the Bay Area—and they can find life ideal in this "middle landscape," locally shaped by viticulture but with all of the modern conveniences of urban life nearby. Among recent buyers of Napa real estate, retired quarterback Joe Montana comes to mind.

Aside from those who own the vineyards and wineries, those who toil among the vines, and those who have taken up residence for the purpose of either commuting to work elsewhere (an increasing phenomenon in much American wine country that lies within commuting distance of nearby metropolitan areas) or retiring, there is a range of other people as well. There are those who own or work in restaurants, motels, shops, light manufacturing, schools, local governments, and transportation facilities, and they have found life in wine country worth staying for. We can find people in wine country who have changed careers rather than leave the area when their employer decided to relocate them in a larger city or a distant state.

Wine Regions and Their Towns

As mentioned already, most of the towns in wine country are small agricultural service centers. They are, of course, also residence places for those who live and work in the area, most of whom neither live nor work on the land. These towns are also the places where the young grow up, are educated, and as often as not, find their own niche in the local working landscape. A number of wine makers began as gofers at local wineries, learning on the job and perhaps taking enology and viticulture classes during the off-season.

Many local wine country towns have a number of wine-related businesses. For example, the city of Napa is home not only to the famous wine train but also to businesses that import and sell wine corks and bottles, make and sell wine seals, design and install pumps and irrigation systems, and provide many other of the needs of local wineries. Stakes, poles, and wires for trellis systems are also sold (and sometimes assembled) by local businesses. In addition, specialized legal services (to deal with the BATF and its maze of regulations, for example) can be found in many wine towns, along with realtors who specialize in sales of vineyard and winery properties, consultants of every sort (to wine makers, to viticulturists, to local nurseries that grow and

sell rootstocks), and distributors who consolidate distribution functions for many smaller wineries.

Aside from businesses that cater directly to the wine industry, there
are growing numbers of businesses that thrive on the tourist trade in
wine country. Restaurants and lodging establishments are the most
obvious, of course, but boutique shops and even mini–outlet malls are
appearing; bus and limousine tours of the vines and hot-air balloon
rides are thriving as well, from California to New York. The city of
Napa, for example, has its own outlet mall, and there are even a few
outlet stores just outside St. Helena. Although many such features of
small towns in wine country in the Napa Valley and Finger Lakes regions were discussed previously, a few additional examples seem
worth mentioning at this point.

Sonoma County, California

There are several splendid winescapes in this delightful rural California county, and some of the more important local service centers include Sonoma and Healdsburg. With the Sebastiani Winery within
its borders and a host of wineries and vineyards nearby (mainly in the
cool Los Carneros AVA), the small town of Sonoma (8,900) is a mecca
for wine country visitors and sits at the crossroads between the Napa
and Sonoma Valleys (which are connected by Highway 12). The
Sonoma mission (Mission San Francisco Solano, the last of California's twenty-one missions to be built) still sets the tone for the town,
which reflects its Spanish heritage in much local adobe architecture.
The old plaza remains the town's central feature, a prominent place to
relax a bit and enjoy the surroundings. Around the plaza are not only
some architectural gems such as the Toscano Hotel (1852) but also
such modern-day attractions as the Sonoma Cheese Factory (a perfect
place for assembling picnic ingredients before you head off to visit
some of the local wineries) and a wonderful French bakery. The Vintners Inn (with its deluxe European-style accommodations) is one example of the use of wine country names. Others include the following: Vineburg Country Market, Vintage Builders, Vintage House
Senior Center, Vintage Medical Inc., Wine Country Balance, Wine
Country Gifts, and The Winemakers Restaurant. The town even has
a Cabernet Building.

Healdsburg (9,800), located along the Russian River at the south
end of the Alexander Valley AVA, is in many ways reminiscent of

Sonoma. The town's plaza, though smaller than Sonoma's, is still its central feature; it is surrounded by restaurants, a bakery, a delicatessen, boutique shops, and a bookstore. Within sight of the plaza is the town's brewpub, the Bear Republic Brewery. Like Sonoma, Healdsburg has wineries both within its border and in virtually every direction outside of town. It makes a superb central stop for visits to the Russian River Valley, Dry Creek Valley, Alexander Valley, and Knights Valley AVAs, a schedule full enough to keep the most dedicated wine country enthusiast busy for days. Local names that reflect viticultural themes in Healdsburg include the following: The Grape Leaf Inn, Grapevine Express, Petite Vines, Vine Street Gas Station, Vineyard Electronics, Vineyard Plaza (a relatively new minimall), Vintage Antiques Etc., Vintage Floors, Vintage II Gasoline, and Wine Country Radio. An added attraction for wine enthusiasts (especially if the weather turns bad) is the Northern Sonoma County Wine Library.

Smaller towns, many no more than hamlets, dot Sonoma wine country as well. Included among them are Kenwood, Glen Ellen (where the Benziger Winery offers a splendid and thoroughly enlightening tram tour through its hillside vineyards), Sebastopol (known more for its Gravenstein apples, as suggested by its Gravenstein Union School District, for example), and Geyserville (hinting of the area's volcanic heritage). More varied and less closely attached to wine country are Santa Rosa (128,000), the county's largest city; Rohnert Park (39,000), home of Sonoma State College; and Petaluma (49,000), a gateway to Sonoma County along Highway 101.

At the southern end of Sonoma County, overlapping with Napa County to the east, is the Los Carneros AVA, located in one of the state's coolest growing regions. Cool marine air from San Pablo Bay bathes the region throughout summer, especially during night and morning hours. Pinot Noir and Chardonnay do especially well in this Region I mesoclimate, and several sparkling wine firms have located here in recent years, including Gloria Ferrer.

Northwestward from the town of Sonoma, the Sonoma Valley lies parallel to the Napa Valley, separated from it by the Mayacamas Mountains. The Sonoma Valley AVA includes not only this narrow valley but also much of the surrounding mountainous terrain on both sides of it. At the south end, Pinot Noir and Chardonnay dominate and to the north, Cabernet Sauvignon, Merlot, and even Zinfandel take over.

Farther north, near Healdsburg, the Alexander and Dry Creek Valleys, each with its own AVA designation, lie parallel to each other, separated only by low hills, and stretch northwestward for several miles. The Dry Creek Valley is especially well known for its Zinfandels, whereas Cabernet Sauvignon in the south and Chardonnay in the north are mainstays in the Alexander Valley. Each valley has more than 5,000 acres of vines and twenty or more wineries. Other AVAs within the county include the Russian River Valley, Chalk Hill, Sonoma Mountain, and Knights Valley.

Temecula, California

Located south of the Los Angeles metropolitan area, yet comfortably north of San Diego, Temecula has become southern California's most important wine-growing region. The Temecula AVA contains some 100,000 acres of land, though there are currently less than 2,000 acres of vines and prospects for expansion seem limited. However, proximity to the densely populated Los Angeles Basin has provided both markets for Temecula wines and a steady stream of visitors to local wineries.

The town of Temecula is somewhat unusual, part preserved "Old Town" and part yielding to shopping centers and suburbanization as the Rancho California development expands. Housing, much of it new, much of it given something of a Mediterranean look (red tile roofs, pastel colors), has been steadily encroaching on the vineyards, most of which are located along Rancho California Road. There are about a dozen wineries in the AVA.

Both the town of Temecula and the surrounding region increasingly receive direct economic benefits from encouraging wine country tourism—a "clean" industry in an area that wants to maintain at least a modicum of its rural charm. The region's wineries all welcome tourists; many have picnic facilities. The annual Temecula Valley Balloon and Wine Festival, which now attracts thousands and lasts for an entire weekend, is held each April; wine, food, and entertainment (including an evening "balloon glow") provide something for everyone. Paradoxically, the very same fogs that slip in through the Rainbow Gap and cool the vines, allowing the production of premium wine grapes in the area, have led to cancellation of balloon launches on several occasions.

Although grapevines have only been growing in this region for about three decades, viticultural names are occasionally seen around the valley. Not far from the Callaway Vineyard and Winery, for example, there is a housing development called Chardonnay Hills. Nearby, another housing development is called Vintage Hills. Columbo's Vineyard Cafe offers dining and catering, and diners can go to the Baily Wine Country Cafe or Café Champagne as well. Aside from wines and viticultural influences, Temecula is also home to the Blind Pig Brewing Co.

Yamhill County, Oregon

Southwest of Portland, west of Interstate 5, Yamhill County lies in the heart of Oregon's Willamette Valley, which is Oregon's coolest (Region I) wine-growing region. Pinot Noir and Chardonnay dominate vineyard acreage here, along with a smattering of Riesling, Gewürztraminer, and even Müller-Thurgau, a hybrid that is grown in large amounts in Germany. McMinnville (17,900) is the major service center for this segment of wine country (and now home as well to the famous Spruce Goose, an airplane once owned by Howard Hughes). An annual Turkey-Rama (barbecue and fair) celebrates the local agricultural bounty each summer. Just outside of McMinnville is the Tanger Factory Outlet Center (its presence reflecting a growing trend in American retailing).

Although farmers in the northern Willamette Valley have not been growing grapes for very long, a few viticultural names have already worked their way into the local landscape. Around Yamhill County, you can find at least the following: Best Western Vineyard Inn Motel, Wine Country Farm (a bed-and-breakfast), Vintage Villa Motor Company and Self Storage, Grape Escape Winery Tours (offering intimate van tours), the Oregon Wine Country (a wine club that specializes in Oregon wines), and Old Noah's Wine Cellar.

Smaller central places around McMinnville include the following: Amity (which gives its name to Amity Vineyards), Lafayette, Dundee, Carlton, Yamhill, and Gaston. Not far away is Portland (440,000, with another million in the surrounding metropolitan area), Oregon's largest city. Straddling the Willamette River (and connected by numerous bridges), Portland is one of America's most enjoyable metropolitan areas.

Yakima Valley, Washington

The Yakima Valley wine region (home to nearly two dozen wineries and many of Washington's finest vineyards) extends along Interstate 82 from just east of Benton City to the area around Zillah; Interstate 82 continues on northwestward to the city of Yakima (55,000). Yakima, by far the region's largest city, is not really within Washington's wine country. However, it is a haven for travelers (including wine country visitors) and offers some wine-related names, including The Vineyards of Yakima Restaurant (which features wines of the Yakima Valley) and Grant's Brewery Pub. Each February, the city hosts the Celebration of Chocolate and Red Wine, a winning valentine combination.

The real wine country service centers, however, are much smaller, and they offer fewer services for visitors. Eastward from Zillah, there are Granger, Sunnyside, Grandview, Prosser, and Benton City. Grandview (7,500) gets its name from the spectacular Cascade volcanoes that can be seen nearby—Mount Rainier and Mount Adams—which tower above the landscape like snowcapped sentinels. Grandview is surrounded by agriculture, from apples and asparagus to peaches and grapes; more important for us, it is the home of the Yakima Valley Wine Growers Association and of Chateau Ste. Michelle, the state's oldest wine-making facility (founded immediately after the end of Prohibition). Between Grandview and Prosser, the old Yakima Valley Highway is now known as Wine Country Road. Each September, the town hosts its Grandview Grape Stomp.

Prosser (4,500) is located just north of the treeless Horse Heaven Hills; both vineyards and wineries are more conspicuous here than in Grandview. The Wine Country Inn (located, aptly enough, on Wine Country Road) is an appealing bed-and-breakfast. The town also hosts the annual Prosser Wine and Food Fair, a celebration of local wines and diverse agricultural products that is held each August, and The Great Prosser Balloon Rally and Harvest Festival.

Zillah (1,900) is one of the smaller agricultural service centers in the Yakima Valley, but it is located in the heart of the valley's orchard and vineyard landscape. Visitors will find that Vintage Road leads to nearby Covey Run Vintners, and the Vintage Valley Parkway will get you to the Zillah Oakes Winery.

Long Island, New York

At its eastern end, Long Island splits into two tails (following the rocky rubble of two parallel glacial moraines), spreading north and south; in both areas, potato fields and other land uses have gradually given way over the last two decades to expanding acreages of grapevines, though there are still more acres of potatoes than grapes. Dotted with wineries as well as vines, these emerging viticultural landscapes are readily accessible to New Yorkers via the Long Island Expressway. There are several small service centers here, of which Cutchogue (2,600) is perhaps the one most closely related to surrounding vineyards and wineries. Long Island has historically been vacation country for weary New Yorkers; the wine industry, still in its youth, has not yet given many names to the landscape, though it has provided one more reason for people to visit the area.

9

$\mathscr{C}$ommunicating About Grapes and Wines

$\mathscr{G}$EOGRAPHERS HAVE ALWAYS BEEN INTERESTED in communications networks because they serve to connect places together, and they represent an important form of spatial interaction. Diffusion studies, for example, have shown how communication systems affect the geography of innovation acceptance. Even though communications about grapes and wines may seem mundane, they are a measure of the strength of consumer interest in wines and the importance of the wine industry, both in America and elsewhere; they are another dimension of viticulture and wine, separate from cultural landscapes but worthy of consideration as an additional dimension of the geography of viticulture. Just as grapes move to nearby wineries and then, once they are made into wine and put into bottles, move through distribution systems toward metropolitan markets, communications about wines, especially those with a consumer orientation, tend to move in the opposite direction—from major metropolitan areas, especially New York, outward. Other such communications, of course, originate in academia, especially in universities that have departments of enology and viticulture.

The existence of the Academy of Wine Communications, headquartered in Napa, gives added credence to the importance of wine communications, both for the industry and the consumer. Harvey Posert, current president of the academy, notes that the academy was formed to support the more than one thousand people in the United States who write about wine (including wine writers, vintners, and winery public relations people). As will soon become apparent, wine is written about in a variety of publications and by writers with many different qualifications and backgrounds. Some are journalists, some academics, and some no more than kindred souls who have found wine an interesting topic to expound upon. Differences in credibility abound.

Numerous books about grapes and wines are available, including both those for viticultural specialists and those for a more general audience interested in grapes and wines. Their numbers grow steadily. Books, however, are not the focus of this chapter; rather, the focus herein is on periodicals and on the rapidly expanding volume of wine information that is available on the World Wide Web. Many books are cited in this book's reference section for those interested. The following list of sources clearly demonstrates the degree to which wines and vines have an impact far beyond molding the winescapes in which they are grown and produced.

Wine in Print: From Academic Journals to Consumer Magazines

Much is regularly published about the wine industry and its products, and specialized publications are designed to reach very different audiences—academia (with its scholarly research on grapes and wines), people within the wine industry (trade publications, common for most American businesses), and wine consumers. In addition, many publications carry articles on grapes or wines as part of a broader spectrum of topics such as food or agriculture.

This section presents an array of such periodicals. Because some periodicals are hard to classify, they have been included in the most appropriate of the three categories discussed, though the following categories are hardly mutually exclusive: academic journals, trade publications, and consumer magazines. The concentration here is on American publications, though a few foreign ones are mentioned because of their international importance in the world of wines and vines.

Academic Journals

Scholarly articles in academic journals, though not generally written for lay audiences, provide the results of current research that is being conducted in departments of viticulture and enology, as well as research results from other disciplines and academic departments on occasion. This research most assuredly affects what goes on in the vineyards and wineries of America, gradually reshaping viticultural landscapes (as with the changes in vine spacing, trellising, and pruning throughout California's wine country or the introduction of new rootstocks) and the wines that are made within them. Furthermore, most academic journals publish book reviews and sometimes brief notes on ongoing research projects as well.

Journals Focusing on Viticulture or Enology

The premier scholarly grape and wine periodical in this country is the *American Journal of Enology and Viticulture*, which covers important research on these topics both here and abroad. Most articles are written by scholars in viticulture and enology programs such as those at the University of California at Davis, the California State University at Fresno, and Cornell University.

A relatively new (beginning in 1990) addition to the academic literature is the *Journal of Wine Research*, published in England by the Institute of Masters of Wine. It is both international in scope and multidisciplinary; its articles, though scholarly in nature, come not only from viticulture and enology programs but also from fields as diverse as economics, geography, anthropology, psychology, and history. Some excellent literature reviews have appeared within its pages as well, providing wine enthusiasts with a considerable treasure of past and present knowledge about wines, vineyards, and even, occasionally, viticultural landscapes.

Another useful publication (from Germany, but written in English) is *Vitis: Viticulture and Enology Abstracts*, which presents abstracts of articles from about four hundred publications. It provides readers with a considerable selection of articles from around the world. Two other foreign academic journals (not in English) are also worth mentioning: *Journal International des Sciences de la Vigne et du Vin* (from France; one of the world's premier journals about wines and vines) and *Journal of the Institute of Enology and Viticulture Yamanashi University* (from Japan; in Japanese except for the title).

Journals with Occasional Articles on Viticulture or Enology

Academia is awash with journals, and because wine and viticulture are topics of interest to researchers in various fields, it is possible here only to suggest some journals in which wine articles appear.

Journals with research articles related to viticulture include the *Journal of the American Society of Horticultural Science, HortScience, and HortTechnology*. Geographic articles on viticulture have appeared in the *Professional Geographer, Annals of the Association of American Geographers*, and the *Geographical Review*.

Scientific articles on wine and related research also appear occasionally in journals such as *Applied and Environmental Microbiology, Journal of Agricultural and Food Chemistry, Journal of AOAC International, Journal of Food Science*, and *Microbiological Reviews*.

Most articles in academic journals are based on field tests, other types of collected data, and scientific analyses; although perhaps less readable than the trade or consumer-oriented publications discussed further on, they are also less likely to be influenced by subjective decisionmaking or by outside influences, whether wineries or retailers. Much of what is written in the more consumer-oriented publications is more subjective—taste, after all, is not easily quantified.

Trade Publications

Most trade publications are produced for people in the wine business—from winegrowers to distributors and retailers—but these publications often have broader appeal as well, especially for wine lovers who want to know more but are unwilling to till the ground covered in academic journals. There is considerable crossover among these journals, with articles written both by scholars and by others either in the wine trade or interested in some aspect of it. Some of these publications provide news items about the industry, as well as wine and grape statistics, statistics on wine sales, data on imports and exports, and a variety of other topics of broad interest. Trade publications are as varied as are the wines that are written about within them.

Among American wine trade publications, the best example is *Wines and Vines*, published in San Rafael, California, by the Hiaring Company (which also publishes an annual *Buyer's Guide* that is an indispensable source of information on the wine industry in the United States). Published monthly, this magazine features everything from letters from readers to feature articles; particularly valuable are special issues that focus on wine technology, vineyards, wine marketing, statistics (one to look forward to every July), management, and several other aspects of the American wine industry. This is one trade publication that has many interesting articles for wine consumers, including short notes on changing personalities in the wine business.

American Vineyard is a relatively young publication (1992), published in Clovis, California. Articles are written both by regular staff members and by viticulturists; coverage is slanted heavily toward California. Also from California comes another useful periodical, *Practical Winery and Vineyard*. It also concentrates more on California, but it does cover broader issues that affect viticulture around the nation (and even, occasionally, internationally). The *Grape Grower*, also more of local interest, is published in Fresno by the Western Agricultural Publishing Company (which also publishes periodicals on cotton, nuts, and other agricultural commodities).

The Wine Institute in San Francisco produces several informative publications, including *Economic Research Report*, *News Briefs*, and *News*. Because of financial difficulties, the institute no longer has an industry statistics publication; however, much of that data is now being provided by Steve Barsby and Associates in a publication called *USWineStats*, published in Oregon. Statistics, along with short articles, can be found in *Impact* (which covers wine, beer, and spirits). Some grape statistics can also be found in *California Fruit and Nut Review*. Various states also collect and publish grape statistics (California's Department of Agriculture, for example, publishes two volumes annually, *California Grape Acreage* and *Grape Crush Report*).

Several regional wine publications provide useful information on the wine industry in the United States, as do other publications that focus more on business aspects of the industry. The following are of interest: *Vineyard and Winery Management* (lots of practical advice for winegrowers), *Western Fruit Grower* (which has occasional articles on viticulture), the *Wine News*, *Wine Business Insider*, *Wine Business Monthly*, the *Massachusetts Beverage Price Journal* (often used for price comparisons over time), *Wine East* (which focuses on the northeastern wine region in the United States), *Market Watch*, and *Central Valley Farmer* (which frequently carries articles on viticulture in California's Great Central Valley).

There are also a few university publications that are of general interest for both the wine trade and wine consumers. *Update* (from California State University, Fresno), *Viticulture and Enology Briefs* (from the University of California at Davis), and two more general publications from the University of California—*Vineyard Views* and *Sustainable Agriculture* (which includes frequent viticultural articles)—are examples. These are informal newsletters, but they often cite, and even summarize, books and studies that have been done recently.

Finally, a few foreign trade publications are also useful, including the following: the *Australian and New Zealand Wine Industry Journal*, the *Australian Grapegrower and Winemaker*, *Technical Review* (from the Australian Wine Research Institute), and *Vini d'Italia*.

Consumer Periodicals

Consumer-oriented magazines and other periodicals that deal with wine (and to some extent with grapes as well) are both numerous and disparate; they run the gamut from classy, even flashy, national publications to newsletters published by either individuals or wineries. In addition, wine articles are featured regularly in many other magazines and in major newspapers as well (although the latter are not considered here, they are sometimes well-known locally).

Wine Periodicals: National to Regional

Two major national wine periodicals stand out above the rest, *Wine Spectator* and *Wine Enthusiast*. They are staunch business rivals, each seeking to win friends among wine consumers, but they have much in common. Both are attractive, large-format (10 by 13 inches) publications with glossy covers, lots of quality color photographs inside, staffs of well-known and respected wine writers who contribute regular columns, and a generous section on wine-tasting results (both publications use a 100-point scale for scoring). Both periodicals are published in New York (leading to occasional charges of neglect by some Californians), both run 80 to 120 or more pages per issue, and both cover the entire world of wines and vines and contain useful information for wine aficionados.

Regular wine columnists for *Wine Spectator* are James Laube, James Suckling, Jancis Robinson, and Matt Kramer; regulars for *Wine Enthusiast* are Alexis Bespalof, Steve Heimoff, Jerry Mead, and Mark Golodetz. Both publications feature guest columnists as well, along with regular features on restaurants; they even provide recipes, with accompanying wine suggestions.

As noted, both publications offer wine-tasting results (done by panels, not individuals—*Wine Enthusiast* even uses an outside judging organization, the Beverage Testing Institute, Inc., in order to maintain objectivity). Tasting results can be useful buying guides, but more important, they can be instructive about the wine industry. For example, in the best of all possible worlds (in vinous terms at least) you would expect to find a strong positive correlation between price (with retail as a standard) and quality (as measured by the panels' ratings)—put simply, you would expect to get what you pay for.

However, tasting results reported in both of these publications clearly show that the wine world is rife with imperfections, that many wines are not worth what you have to pay for them. Although there is a general tendency for the highest priced wines to score high, it is not uncommon to find wines that sell for much more than their scores would warrant. As eminent wine writer and historian Leon Adams once put it, "As with anything else you buy, you want value for what you pay. I therefore should tell you that the principal art of the vintner is in persuading you to pay more for wine than it is worth."

In no way is this meant to be a criticism of either the publications or the 100-point scale (which has both advocates and opponents)—on the contrary, tasting panels provide us with lots of food for thought. Another caveat, of course, is that your taste might differ considerably from a panel's, so you should do your own sampling as well. Keep in mind that wine tasting (even when it is given the authoritative appearance of a 100-point scale) is ultimately subjective. If you look at enough different wine ratings, then you will see how much they can vary when they rate exactly the same wines, so don't feel put off when you do not agree with someone's rating of a particular wine. If "experts" cannot agree, then why should you?

Published now in Carson City, Nevada (after moving from California, which is a story in itself, told at some length in one issue), Jerry Mead's *Wine Trader* is the most iconoclastic of American wine publications. It is not as glitzy in appearance as some wine magazines, but *Wine Trader* has modernized considerably from its earlier versions, most recently adding a colorful, glossy cover. Most articles and commentaries focus on American wines and wine regions, with a strong emphasis on California. There are several columns of interest, including a "Health and Social Issues Report," by Elisabeth Holmgren (from the Wine Institute), and a feature about wine information on the Internet by Jim Wallace.

From Massachusetts comes the *Quarterly Review of Wines*, which tends to have a strong international component, though American wines and wineries are not neglected. *Wine and Spirits*, published in Princeton, New Jersey, offers a reasonable balance between articles on national and international wines. From South Pasadena, California, comes the *Wine Journal*, which, despite its California location, has an outlook and coverage that are thoroughly international. The *Wine News* is published in Coral Gables, Florida, and is also international in scope. Arriving from across the Atlantic, *Decanter* is another high-quality wine publication aimed at the wine-consuming public; it is heavy on international coverage. A relative newcomer, aimed at a younger audience (Generation X, as the media has unfortunately dubbed it), is *Wines International*, published in Santa Rosa, California.

Regional wine publications originate in scattered locales, especially in and around wine country; some are published annually, others quarterly or even monthly. Many are touristy. They include lists of current events (including many winery events), information about local wineries and restaurants, maps (often neither to scale nor accurate), and some feature articles on wines, wine makers, or viticulture. Their numbers are legion (finite, but uncountable—I think that is how mathematicians might phrase it!). Good examples include the following: *California Visitors Review* (published in El Verano, California, just outside the town of Sonoma), *Napa Valley* (published in Redwood City, California), *Steppin' Out: California's Wine Country Magazine* (published in Fort Bragg, California), *Guest West* (published in Napa, California), *Wine Country Almanac* (published in Paso Robles, California), *Amador County Wine Country* (published in Jackson, California, by the Amador Vintners Association), *Sierra Foothills Vine Times* (published in Placerville, California), *Discover Oregon Wineries* (published in Portland by the Oregon Winegrowers' Association and the Oregon Wine Advisory Board), *Latitude 46* (published in Seattle by the Washington Wine Commission), the *Grape Vine* (published in Prosser, Washington, and providing information on the wines and wineries of the Yakima, Columbia, and Walla Walla Valleys), *Colorado Wine Country Tours* (published in Grand Junction by the Colorado Wine Industry Development Board), *Wines of the Texas High Plains* (published in Lubbock by the Texas Wine Marketing Research Institute), *Virginia Wineries: Virginia Wine Country* (published in Richmond by the Virginia Wine Marketing Program), *North East: Welcome to Wine Country* (published in North East, Pennsylvania), *Finger Lakes Region Travel Guide* (published in Penn Yan, New York), *Cayuga Wine Trail* (published in Fayette, New York), and the *New England Wine Gazette*. Almost anywhere that you find vineyards and wineries you can find someone writing about the area (check with local chambers of commerce or with wineries in a particular area).

Food Publications Regularly Including Wine Articles

The marriage of food and wine, perhaps as old as winegrowing itself, finds its way into the publication world in a significant way. Many food (or food and wine) magazines feature at least a wine column and often feature menus that include wine as well.

Food and Wine is one good example of this genre. Published in New York, this magazine regularly publishes wine articles. Contributing editors include Robert Parker, Stephen Tanzer, Elin McCoy, and

John Frederick Walker. *Gourmet*, also published in New York, is another. Wine writer Gerald Asher contributes a regular column titled "Wine Journal," which is usually an in-depth look at some wine, wine region, or facet of the wine industry. Two others round out the national list: *Cook's Illustrated* (published in Brookline, Massachusetts) has a wine article in each issue, and *Bon Appétit* (published in Los Angeles) features a monthly column, "Wine and Spirits," by wine writer Anthony Dias Blue. Also out West, wine writer Bob Thompson contributes an all-too-brief monthly column to *Sunset*. Even more health-oriented food publications are recognizing the value of wine as part of a healthy diet. *Eating Well*, for example, maintains on its staff a wine consultant, Irving Shelby Smith, who regularly writes wine articles for that magazine.

Newsletters from Wineries, Wine Shops, and Other Sources

Winery newsletters have increased rapidly in number. Although their content runs the gamut of topics and some of the writing is downright self-serving, interesting tidbits appear in these newsletters frequently enough to make browsing through them worthwhile. Particularly interesting are discussions of seasonal weather patterns, new techniques and technologies that are being explored, and brief articles by wine makers and viticulturists—those who do the real work, yet are seldom heard from. Many of these newsletters are available on the World Wide Web as well.

Of particular interest are newsletters from wineries (most available free of charge), including "Arrowood 'Vignettes'" (Arrowood Vineyards and Winery, Glen Ellen, California), "Belvedere Exchange" (Belvedere Winery, Healdsburg, California), "Bridgeview Vineyards Newsletter" (Bridgeview Winery, Cave Junction, Oregon), "Chandon Club News" (Domaine Chandon, Yountville, California), "Clos Pegase Newsletter" (Clos Pegase Winery, Calistoga, California), "David Bruce Newsletter" (David Bruce Winery, Los Gatos, California), "Dry Creek Vineyard Gazette" (Dry Creek Vineyard, Healdsburg, California), "Eberle" (Eberle Winery, Paso Robles, California), "Foppiano Grape Tidings" (Foppiano Vineyards, Healdsburg, California), "Handley Happenings" (Handley Cellars, Philo, California), "In Vino Veritas" (Renwood Winery, Plymouth, California), "Mount Palomar Winery News" (Mount Palomar Winery, Temecula, California), "Over a Barrel" (Hyatt Vineyards Winery, Zillah, Washington), "Phelps News Briefs" (Joseph Phelps Vineyards, St. Helena, California), "Pressing News" (Thomas Fogarty Winery, Portola Valley, California), "Simi Winery Newsletter" (Simi Winery, Healdsburg, California), "St. Clement Connection" (St. Clement Vineyards, St. Helena, California), "St. Supéry Chronicles" (St. Supéry Vineyards and Winery, Rutherford, California), "Storrs" (Storrs Winery, Santa Cruz, California), "Sutter Home Newsletter" (Sutter Home Winery, St. Helena, California), "The Pressing Issue" (Trentadue Winery, Geyserville, California), "Trumpet of the Vines" (Ferrari-Carano Winery, Healdsburg, California), "Baily and Temecula Crest Wine and Food News" (Baily Vineyard and Winery, Temecula, California), "Tualatin Notes" (Tualatin Vineyards, Forest Grove, Oregon), "Views from the Valley," (Orfila Vineyards, Escondido, California), and "Uncorked" (Washington Hills Cellars, Sunnyside, Washington). Ask your favorite wineries about newsletters or home pages on the Internet.

Wineshops have also begun publishing newsletters that are often quite informative. Although they often tell you about wines that they are selling, you can glean much from them about everything from the quality of different vintages to the tales of individual winegrowers and the market conditions for wines. Personal idiosyncrasies abound in some, but always, it seems, with the intention of providing wine consumers with usable information. Among the best of this genre are the following: "Wine Club Newsletter" (Santa Ana, San Francisco, and Santa Clara, California), "Stock Report" (Wine Exchange, Orange, California), "The Wine Country" (The Wine Country, Signal Hill, California), and "Hi-Time Wine Cellars" (Hi-Time Wine Cellars, Costa Mesa, California). The following newsletters, which are less extensive, are available from these wineshops: Solano Cellars, Albany, California; Northridge Hills Liquor and Wine Warehouse, Northridge, California; Jensen's, Palm Springs, California; Wine Cask, Santa Barbara, California; St. Helena Wine Center, St. Helena, California; Los Angeles Wine Co., Los Angeles and Palm Springs, California; The Wine Cellar, Rohnert Park, California; Wine Ventures, Tenafly, New Jersey; Park Avenue Liquor Shop, New York, New York; and Duke of Bourbon, Canoga Park, California.

A variety of other wine-oriented newsletters is published, both by individuals and by organizations. Some are free, others are relatively cheap, and a select few are rather expensive. Some of the better known (or in some cases just plain interesting, if not so well known) newsletters in this category include the following: "The Wine Advocate" (published by Robert Parker and considered by many to be the most influential of all wine publications), "Connoisseur's Guide to California Wine" (published in Alameda, California), "The Fine Wine Review," (published in San Francisco by Claude Kolm, who has gained a reputation for reliable advice), "The Vine" (published in England by Clive Coats), "The Wayward Tendrils Newsletter" (published by the Wine Book Collector's Club, Santa Rosa, California, and full of interesting notes and essays about wine books, including reviews of recent books and discussions of old-timers), "The Baxevanis American Wine Review" (published by John Baxevanis, a professional geographer and wine writer, Stroudsburg, Pennsylvania), "Fruit Winemaking Quarterly" (published in Sebastopol, California), "California Grapevine" (published in San Diego, California), "Florida Wine Bulletin" (published in Coral Springs, Florida), "International Wine Cellar" (published in Cherokee, New York), "Macs Wine Fax" (published in Pleasanton, California), and "The Petite Sirah Report" (published in Tiburon, California).

Wine-Buying Clubs

For a geographer, these are a particularly puzzling group because many do not identify in their advertisements where they are located. Rather than trace them down, I've left a bit of mystery for you. They are relatively new on the wine scene, their wine offerings range from international to local, and they ship directly to consumers. Examples of this somewhat different way of distributing the fruits of the vine (offered with no endorsement) include the following: Wine of the Month Club (1-800-949-WINE), Gold Medal Wine Club (1-800-266-8888), The California Wine Club (1-800-777-4443), Oregon Pinot Noir Club (1-800-847-4474), California Wine Tasters (1-800-941-WINE), A Taste of Ambrosia (1-800-435-2225), The International Wine Society (1-800-430-WINE), New England Wines (1-800-WINE199), California Winemakers Guild (1-800-858-WINE), Chateau Select Wines (1-800-820-7793), Wine Finders (1-800-845-8896), Prince Michel de Virginia Tastevin Wine Club (1-800-549-7372), and the Oregon Wine Club (1-800-WINECLB).

Wine on the World Wide Web

Worthwhile advice and information about wine and the Web can be found in Jim Wallace's "On-Line Report" in the *Wine Trader*. From ANSI (American National Standards Institute) to URL (Uniform Resource Locator, which refers to an Internet address), Jim provides translations and pointers for those not yet adept with computers and the arcane vocabulary that accompanies them (which can rival the government's acronym list). Wine sites are being added rapidly to the Web. As you might expect, major pages on the Web have been established by leading consumer-oriented wine publications. The *Wine Spectator* (http://www.winespectator.com) and *Wine Enthusiast* (http://www.wineenthusiastmag.com) sites provide wine lovers with reams of information, from updates on wine happenings and wine reviews to connections with myriad wine-related sites.

Among the most commonly visited wine sites is "American Wine on the Web," which offers a host of visits to different sites with no more than a click of the mouse. One example is "Cowper's Wine Glossary," where wine maker Etienne Cowper (currently at Mount Palomar Winery in Temecula, California) has assembled notes on nearly three dozen different cultivars, from Aurore to Zinfandel. These notes describe uses, flavors, and other characteristics of each grape that might be of interest to wine lovers. Each "issue" of "American Wine on the Web" offers a table of contents, editorial comments, feature articles, regional reports, and a food center entry. Managing editor Richard Jones, for example, provided an enlightened editorial, "The Terroir Tempest," not long ago, and another regular, Tom Marquardt, did a varietal report, "Exploring Sangiovese in America." Many wine writers offer their own Web pages.

If you're thinking about visiting wine country, check it out on the Web—you are likely to find everything from maps and descriptions of wineries to suggested lists of local hostelries and restaurants. For example, Richard Leahy's "Regional Guide to Virginia Wineries" offers a look at five wine regions within Virginia, including accompanying maps. You can then click on any one of the five regions, where you will find a list of wineries (forty at last count, but watch for more in this emerging wine-growing region). The Napa Valley has a Web site called "Welcome to the Napa Valley" that will tell you all you need to know before you visit, from climate data to churches, from sightseeing activities to restaurants; there is even a bulletin board on which you are invited to post your favorite non–Napa Valley wine. Maps show winery locations, and you can even click on the Chardonnay Golf Club.

"Virtual visits" to many wineries are available. One of the most elaborate so far is offered by Rutherford Hill Winery, and the "visit" is composed of a full-motion educational video of the winery; you can select everything from pressing the grapes at harvest time to seeing barrels toasted by local coopers.

For those in search of more "scientific" information about wines and viticulture, the Department of Viticulture and Enology at the University of California at Davis, for example, has its own home page. The contents range from an introduction to the department's staff members to a detailed bibliography. If you are wondering about a group known as "The Trellis Alliance," here is the place for you to find out what it is about.

If you are looking for a basic discussion of wines and vines without buying a book, you can print out much of the "Internet Guide to Wine," by Bradford and Dri Brown (who, like many on the Internet, can be contacted via e-mail—these authors can be reached at bradb@netcom.com).

If you want to know about events in American wine country, try "American Wine on the Web," mentioned earlier. On this site, "American Wine Happenings" is published and regularly updated. You can announce events by contacting the Web site by e-mail at winefreak@aol.com. Among the most unusual wine events in one of the recent listings was a peanut butter and wine pairing at Chouinard Vineyards (not far from San Francisco).

10

$\mathscr{A}$merica's Viticultural Future

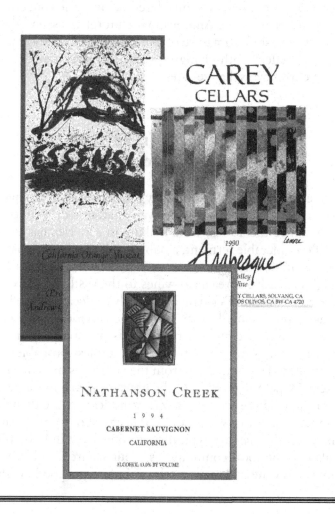

$\mathscr{H}$OWEVER MUCH THEY ARE COUCHED in caveats, utterances about the future must be considered with a healthy dose of skepticism. Inferences about the future of an industry (especially one as geographically diverse as the wine-growing business in the United States), made primarily from trends in recent years, provide only inklings of what could happen, with hints of possibilities, with suggestions about ways that the course of events might be swung in one direction or another.

Aside from social and cultural factors that are likely to influence wine consumption in the United States in the years immediately ahead, two other areas are considered below. One concerns problems that are arising in some American viticultural areas, and the other has to do with the shifting nature of wine imports and exports. The latter, of course, reflect both changing consumer preferences and fluctuations in currency exchange rates.

Cultural Trends and American Wine Consumption

The fate of America's wine industry lies primarily in the hands of wine consumers, not those of winegrowers; patterns of wine consumption are key. Since the 1960s, wine making in America has experienced a time of considerable ferment, a bubbling over of enthusiasm and technological innovation that has stretched from the halls of academe through expansive acreages of vines to the vastly improved quality of wine in the consumer's glass. The last three decades have brought not only an unprecedented expansion of the American wine industry but also a diffusion of that industry from its previous centers—primarily California and New York—to increasingly less probable wine-growing regions in far-flung areas from the volcanic slopes of Maui and the Western Slope of Colorado to the gravelly glacial moraines of eastern Long Island. There are commercial wineries in more than forty states, and most people live within driving distance of some version of wine country, where they are typically welcomed not only by the wineries but also by the local communities. Tourists are a welcome source of income, and wine making is a "clean" industry, a good neighbor.

Recent Trends in Wine Consumption

Despite the best efforts of early American wine enthusiasts such as Thomas Jefferson, Americans have never accepted wine as a typical part of daily living. Like the Australians, we are more likely to be beer drinkers than wine sippers. This does not mean that there are no Americans who regularly enjoy the manifold pleasures of wine—some people, after all, are buying all those American wines, reading all those wine-related publications, and visiting wine country locations across the nation. It is true that in Italy, France, and Spain, wine seems as much a part of the daily table as bread—both are appreciated, both are integral parts of an eating experience that serves to satisfy both body and soul. Nonetheless, per capita wine consumption in those countries has declined in recent decades, and many in those nations drink no wine.

Although recent times have caused considerable enthusiasm and optimism to arise among American winegrowers, the trend in per capita wine consumption in the United States has been heading steadily downward since 1986—not a trend to smile about. This trend is deceptive; it should neither be discarded, nor should we read too much into it yet. It is deceptive because it masks the different consumption trends that are occurring by wine type. Also, though per capita wine consumption in the United States may be low compared to many other countries, total consumption is large simply by virtue of the size of the nation—more than 265 million residents and growing.

Wine-cooler consumption has almost disappeared. Per capita consumption of wine in coolers was just over one-half gallon in 1986 but has fallen to below one-tenth of a gallon; wine in coolers has been replaced primarily by malt-based substitutes. Furthermore, during the ten-year period since 1986, per capita consumption of fortified wines (already low by historical standards) declined by 50 percent and that of sparkling wines declined by about one-third—the only good news is that table wine consumption has fallen much more slowly. At the same time, consumers in recent years have been "trading upward," drinking better quality table wines, though not in larger quantities. There has been both a solid improvement in sales of more expensive table wines and (at least until 1996) a decline in the sales of "jug" wines and "wines-in-a-box."

Sales of American wines have been up in recent years, but so have American wine prices. Consider, for example, wine maker Chuck Wagner's Caymus Cabernet Sauvignon Special Selection (which consistently ends up in *Wine Spectator*'s annual top 100 list): The 1984 vintage had a list price of $38, whereas the 1992 carries a $100 price tag (this and subsequent prices are winery list prices for one 750-milliliter bottle). Yes, you might say, but that wine is as much an exception as it is exceptional; however, in reality this is closer to the rule with respect to wine price increases, at least in percentage terms.

James Laube (1996a:140) began an article on new Zinfandel releases by saying that "about the only sour note for 1994 is that these bright, lush wines are often scarce and rising in price." He provided some examples: Green and Red's Napa Valley Zinfandel went from $8.50 for the 1985 vintage to $16.00 for the 1994; Nalle's Dry Creek Valley Zinfandel went from $8.00 for the 1985 to $16.00 for the 1993; Lytton Springs' Sonoma County Zinfandel (something of a gold standard for California Zinfandels) jumped from $8.00 for the 1985 to $18.00 for the 1994. Furthermore, none of these is near the top of the current Zinfandel price range, which is now hovering close to $30.00. These prices are all for California wines, but other American wine prices are rising as well, though perhaps not quite as fast. Writers Jay Stuller and Glen Martin (1994:37) noted the following, even before the latest round of wine price increases:

> Some vintners ask consumers to pay ungodly prices for their wines. A few surely deserve the premium, for creating rare and wondrous wine that meets the caviar wishes, champagne dreams, and hefty pocketbooks of a very small segment of the public. An allocated supply and high demand for a great wine allows it to command high prices, as is true with any product. But other vintners get $14 for a wine that's not appreciably different from a $10 bottle.

In another article, Jeff Morgan (1996a:10) commented that "the current demand for California wine is outpacing supply, and it doesn't take an economist to figure out that this means consumers will also soon see higher prices for California premium wine." According to the *Wine Spectator*, there are now nine California wines that retail for $70 or more per bottle (with the most expensive being the 1994 Diamond Creek Cabernet Sauvignon at $250).

In the short run, supply and demand for American (and particularly California) wines are certainly out of balance. Strong demand for premium wines seems to be a result of several variables, including such disparate factors as the aging of the "baby boomers," who are beginning to turn fifty now and reaching their peak wage-earning years, and the many studies in recent years that extol the health benefits of moderate wine consumption (especially of red wines, so some say). More will be said about both of these trends in subsequent sections.

On the supply side, Jeff Morgan (1996a:10) has recently noted, "Already, consumer thirst has left some wine makers struggling to find enough grapes." California wine-grape prices have been driven upward in part because more vintners are chasing the same grapes, especially Merlot (which has become the "hot" red wine of the 1990s), Chardonnay, and Cabernet Sauvignon. In recent years, top prices for Napa Valley grapes have soared: Some Chardonnay lots have reached $2,000 per ton, Merlot prices in some cases have reached $2,500 per ton, and the best lots of Cabernet Sauvignon have gone for more than $3,500 per ton. A general rule of thumb suggests that if you divide the grape price per ton by 100, you will get a good estimate of the price per bottle that a wine made from them will have to sell for. In other words, a wine maker who pays $3,000 per ton for Cabernet Sauvignon grapes is going to have to get around $30 per bottle for a wine made entirely from those grapes.

In the short run, California wine-grape supplies are caught in a pincers—on the one hand, there have been smaller than normal crops (especially in 1995, but also in 1994), and on the other hand, there is phylloxera, which has damaged thousands of acres of vines in Napa and Sonoma Counties (and is affecting vineyards elsewhere as well, from Santa Barbara to El Dorado Counties). One option is for wine makers to blend in more grapes from other wine-growing regions in order to cut average costs somewhat; the Lodi AVA, for example, has been supplying more grapes than ever to coastal wineries. Many are doing that, though wine writers are quick to point out the decrease in wine quality that has resulted, especially for Merlots. Another effect of this strategy is a predictable rise in the price of Lodi grapes, though prices remain below those paid for coastal grapes.

Growers are responding to higher prices by planting new vineyards as well, but these will take a few years to come on line. Phylloxera-ridden vineyards are quickly being replaced, for example, and new planting densities and trellising systems are likely to increase yields substantially in the years ahead. At the same time, growers in Lodi and other areas where vineyard land is still available are heavily planting new vineyards, especially with Merlot, Cabernet Sauvignon, and Chardonnay. Shortages of other cultivars, such as Sauvignon Blanc, are likely to be the outcome of current trends, because everyone wants to plant the varieties that are now in high demand. Such patterns—high prices followed by widespread new plantings—have been seen before in the Golden State, where boom-and-bust cycles have been part of the wine industry's legacy.

Although new plantings in the Napa Valley are geared toward producing higher yields, questions do remain, and arguments abound, about whether quality will be maintained—the combination sounds too good to be true. *Wine Spectator*'s Matt Kramer (1996:29), for example, in speaking about wine quality, has argued that "what one wants is density. You can feel it; it's practically a whole-body experience. And that comes, above all, from low yields." Similarly, wine writer Bill St. John (1995:52) wrote that "lower yields mean greater concentration (and more flavor) in wine grapes." Winegrower Justin Meyer (1989:88) informed us that "to maximize profit, a grower wants to maximize crops. Yet he must be careful to avoid 'overcropping.'" Somewhat differently, wine writer Hugh Johnson (1989:33) has pointed out that the French dwell on a rule that suggests that quantity and quality are inversely related. However, he went on to suggest that "today it is more realistic to rewrite the golden rule to read 'higher quantity means lighter, more rapidly maturing wine'— the style which is most in demand." In that same vein James Laube (1995:14) commented, "It's my view that virtually all California wines, regardless of their color, varietal character or history, are best consumed in their youth . . . the trend around the world is to make wines that are more accessible earlier." On a more practical level, the owners of Matanzas Creek Winery, in Sonoma County, found that by using a lyre trellising system they could raise Merlot yields considerably (to 8 tons per acre) but *only* with a significant loss in grape quality, which they decided was unacceptable (after all, they produce the most expensive Merlot in the United States). Will higher grape yields in the Napa Valley bring us higher quality grapes and wines—even though lighter and more approachable—or are they geared more toward increasing winegrowers' profits? Only time will tell, though many wine writers are optimistic. Jeff Morgan (1996a:10), for example, noted that "for consumers, the best news is that new vineyard plantings often use the latest modern viticulture techniques, and the quality of California grapes—and wine—is improving across the board."

In the meantime, winegrowers in other American wine regions are finding new market niches for their wines as California prices move steadily upward. Washington wines have become especially popular because of their combination of relatively lower costs and rising quality. As of 1996, there were a number of very good Washington Merlots available for around $10 per bottle, for example, a price range for that varietal wine that has been under considerable pressure in the Golden State. Wines from New York, Virginia, and even Texas are becoming more competitive.

At the same time, even with the current low level of the dollar abroad compared to its level a decade ago, imports are likely to make inroads under current domestic production conditions. Wines from Chile, Australia, and South Africa are especially attractive, as are French wines from relatively new (and highly improved) regions in the south of that country where new plantings of Cabernet Sauvignon, Merlot, and other cultivars are now making their presence felt. Watch for French wines with appellations such as Vin de Pays de Coteaux de l'Ardèche and Vin de Pays d'Oc.

Wine and Health: A Continuing Debate

Perhaps nothing has been more beneficial to the American wine industry in recent years than Mike Wallace's 1991 segment "The French Paradox" on *60 Minutes*. His central theme was simple: Though the French consume more fats than Americans and smoke more than Americans, they have substantially lower heart attack death rates than Americans. That, of course, is the paradox, and the explanation provided for that paradox by Mike Wallace and those he interviewed was that the French people gain protection from heart disease from their consumption of wine, primarily red wine. Given the American public's love of anything that smacks of being a panacea (we've always been searching for that "fountain of youth"), the immediate results were predictable: Red wine sales in the United States spiked upward the day after the show aired. Subsequently, red wine sales have remained strong *and* more evidence has accrued in support of the health benefits of moderate wine consumption.

It is true that the French have lower heart attack death rates than Americans (and slightly higher life expectancies as well). Whereas heart disease accounts for about 37 percent of all American deaths each year, it accounts for only about 22 percent of all French deaths. Heart attack death rates in one recent year were 113.5 per 100,000 for American men, compared to only 73.2 per 100,000 for French men; the rates in both countries were much lower for females, with rates of 88.9 per 100,000 for American women and 54.8 for French women. Other Mediterranean countries—Spain, Portugal, Italy, and Greece, for example—also have lower heart attack rates than we find in the United States.

Neither Mike Wallace nor most others who have talked about "The French Paradox" have mentioned Japan, however, whose residents have the world's lowest heart attack death rates by a considerable margin (28.2 per 100,000 for males and 20.9 for females). Japan, of course, is hardly a major wine-consuming nation, though it does have a very high per capita rate of alcohol consumption. The Japanese do, however, consume a diet that is much more similar to that of the people around the Mediterranean than to that of the typical American.

Subsequent to the original *60 Minutes* segment, an update was done in late 1995. It further confirmed the beneficial aspects of wine, citing several newer studies that have supported claims about the heart benefits of moderate wine consumption.

Studies in medical journals have included within their samples a diverse range of people as subjects, yet researchers have reached similar conclusions about the health benefits of light-to-moderate alcohol consumption. For example, results from the Framingham Study (focused on residents of Framingham, Massachusetts), one of the most thorough studies of risk factors associated with heart attacks in that it included the relationship between cholesterol and heart disease, along with the results of studies of Japanese men in Hawaii, residents of Busselton in western Australia, and drinkers in Alameda County, California, have led researchers to accept the "U-shaped" alcohol-mortality curve. According to these data, along with data from numerous other studies, life expectancy for teetotalers is somewhat less than that for people who drink up to two drinks per day; beyond that consumption level, however, according to most (though not all) of these studies, life expectancy begins to decline. In 1995, the Copenhagen Heart Study found that wine drinkers, specifically, had greater life expectancies than either teetotalers or imbibers of other types of alcohol. As physician David Whitten (1996a:36) said in a recent *Wine Spectator* article, "This is beginning to be a very old and stable literature, with very secure qualitative results—wine drinkers live longer and stay healthier throughout their lives." In early 1996, the U.S. Department of Agriculture, for the first time ever in its recommended *Dietary Guidelines for Americans*, went so far as to say that moderate alcohol consumption may be beneficial to a person's health.

Physicians David Whitten and Martin Lipp (1994), in their book *To Your Health!*, reviewed numerous studies of alcohol and health, including most of those cited above, and summarized them for lay consumption. Among their positive findings and recommendations, these two physicians informed us (1994:48), "For all the caution of scientific committees and governmental advisory boards, and in spite of the strident rhetoric of the antialcohol community, no other food or beverage comes close to achieving the positive effect of the consumption of light-to-moderate amounts of alcohol, especially wine, on cardiac health . . . the evidence thus far is massively in favor of wine." Whitten and Lipp were careful to emphasize several points; for example, light regular wine consumption is defined as one glass of wine per day for females and two for males, consumed with meals. They also noted, quite forcefully, that there is no benefit to be gained by saving up and having seven to fourteen glasses of wine on Sunday and drinking nothing during the rest of the week. Furthermore, they were careful to note the risks of alcohol abuse and recommended no alcohol at all for those who are abusers or potential abusers.

There is no doubt about the manifold dangers of alcohol abuse, though regular wine drinkers are thought by many to abuse alcohol less than many other groups, especially drinkers who consume primarily spirits ("hard liquor"). A 1993 study, the National Household Survey on Drug Abuse, reported that there were about 103 million people over age twelve in the United States who were currently consuming alcohol, of whom some 11 million were classified as "heavy drinkers," down somewhat from 12 million in a comparable 1985 survey. Furthermore, the economic costs of alcohol abuse—including reduced productivity, mortality losses, and treatment—are estimated to be in the range of $100 billion annually in the United States. Beyond those figures, alcohol abuse takes its emotional and physical toll on families and contributes to everything from homicides to accidents.

One continuing area of disagreement about alcohol consumption concerns women. Whitten and Lipp (1994:92) concluded that "women who enjoy light, regular wine drinking should evaluate any alleged risk of that habit in light of the readily demonstrated health benefits associated with this level of consumption . . . we believe any woman who has shown herself to be a stable light, regular wine drinker, should be allowed to continue that habit without regard to whether she is pregnant or nursing a newborn baby." However, in a recent survey of eminent female physicians that asked them how they advised their patients about wine and health, Susan Lang (1996:72) found that "while these doctors differed on details, they all agreed that there are other, less risky ways to modify heart disease risk." Among them are the following: a low-fat diet, exercise, losing weight, not smoking, and reducing stress. Lang cited several physicians, one of whom coauthored a study of the effects of alcohol on women's mortality, which involved a sample of 86,000 women between the ages of thirty-four and fifty-nine and went on for twelve years. The coauthor of that study, physician JoAnne Manson, was quoted by Lang (1996:72) as follows: "In women under age 50, we saw *no* benefit of alcohol consumption. However, in women 50 and over—who are post-menopausal and thus have a higher baseline risk of heart disease—we began to see a protective association with moderate drinking." She also noted, though, that more than two drinks a day led to other problems, including liver disease, breast cancer, and gastrointestinal bleeding.

Other evidence suggests that even moderate alcohol intake among women may increase the risk of breast cancer. Furthermore, there remains considerable debate about the relationship between even moderate alcohol consumption, pregnancy, and possible harm to the early development of the fetus. In Lang's article, for example, she cited physician Gigi El-Bayouni as follows (1996:74): "Pre-menopausal women have to be especially careful about alcohol. We know that the important time for fetal development is the first two months, when many women don't yet know that they're pregnant." Most of the physicians interviewed by Ms. Lang suggested that three to five glasses of wine per week for most women would not be a problem, and most of them said that they personally consumed wine at about that level. For women, and for that matter for everyone, the benefits of light-to-moderate alcohol consumption must be compared to the risks. With no other changes in lifestyle, however, simply adding a glass or two of red wine to your diet is never going to be a panacea—it will not clear arteries of a lifetime of sludge, erase the potential consequences of a high cholesterol diet, or make up for a sedentary lifestyle.

In 1992, Lewis Perdue's book *The French Paradox and Beyond* followed up on Mike Wallace's initial *60 Minutes* segment. The book ranges widely over the health benefits of wine, but it considers other aspects of lifestyle as well; those interested in a healthier, and possibly longer, life are going to have to do more than wash down hamburgers and fries with a glass of Zinfandel. Lewis Perdue, Keith Marton, and Wells Shoemaker (1992:3–4) cited physician R. Curtis Ellison (chief of preventive medicine and epidemiology at Boston University's School of Medicine and one of the physicians that appeared on *60 Minutes*) as saying that the "French secret" comes down to a combination of wine, food, and lifestyle. Ellison suggested that the regular consumption of wine (especially red wine) with meals, along with fresh fruits and vegetables, slower ingestion of meals and fewer snacks, less red meat, less milk and more cheese, and less butter and more olive oil, combine to provide the French with lives less threatened by heart disease.

This important dietary thread, in which moderate wine consumption *with* meals plays a significant role, has been followed up by both physicians and dietitians, particularly with respect to what has become known as the "Mediterranean Diet." Four years before the first *60 Minutes* broadcast popularized the health benefits of wine, journalists Carol and Malcolm McConnell (1987) wrote about many benefits of the Mediterranean diet. Grapes—eaten raw, as raisins, or in wine—have long been a staple food around the Mediterranean. Grapes and wine then, coupled with other fresh fruits and vegetables, olives and olive oil, and such complex carbohydrates as wheat (in bread, pasta, and couscous) and rice, form the basis of the diet that has been eaten by peoples around the Mediterranean for centuries. Meat has been eaten only sparsely, and milk is consumed foremost in the form of cheese (often grated over pasta and rice dishes for a wonderful combination of flavor and nutritional value that is hard to beat). Throw in a little fish for variety, and you pretty much have a picture of the Mediterranean diet. They informed us (1987:16), with neither the bravado nor the fanfare of Mike Wallace, that "during the course of this research, it has become clear that for millions of people the unique combination of foods in the traditional Mediterranean diet has dramatically reduced the incidence of heart disease and cancer." By that time, the World Health Organization's Expert Committee on

Cardiovascular Disease had also noted the considerable association between longevity and geography (with its dietary implications), mainly that low rates of coronary heart disease were found in people around the Mediterranean Basin and in parts of Asia (especially Japan, as was noted earlier).

Although wine is without doubt the "sexiest" part of the Mediterranean diet to talk about, its health benefits are best viewed within the broader context of dietary and other lifestyle changes. The McConnells, for example, looked at the various benefits of different parts of this diet, including the role of whole grains and cereals, olives and olive oil (a monounsaturated oil that receives a minimum of processing), grapes and wine, fresh vegetables and fruits, fish and fish oil (the heart benefits of omega–3 fatty acids have become better known in recent years), and garlic and onions. That the latter bulbs are both integral parts of Mediterranean cooking and good for you has been borne out by recent research, much of which has focused on the manifold benefits of antioxidants.

Quercetin, for example, an ingredient found in onions, garlic, broccoli, and red wines (but neither in spirits nor beer), has been shown to have significant chemopreventative benefits, especially against cancers such as squamous cell carcinoma. Concluding their discussion of quercetin, physicians Whitten and Lipp (1994:73) noted that "the existence of a potent anticarcinogen in wine, even in small quantities, illustrates once again the 'special' character of wine in the diet as compared with other alcoholic beverages." Maybe we can encourage Mike Wallace and *60 Minutes* to add garlic, onions, and other known beneficial antioxidants to their next update on "The French Paradox."

The Mediterranean diet and its benefits received even greater attention in 1993, when they were the focus of an international conference in Cambridge, Massachusetts, sponsored by the Harvard School of Medicine and Oldways Preservation and Trust. Scientists, scholars, journalists, and others joined together to discuss the health benefits of a Mediterranean lifestyle. Out of the many scholarly and other discussions at that conference came considerable agreement about the health benefits of eating the Mediterranean way.

The conference's sponsors, along with the World Health Organization, subsequently endorsed the Mediterranean Diet Pyramid (Figure 10.1). The essential building blocks of a daily diet are complex carbohydrates—breads, pasta, rice, couscous, polenta, bulgur, other grains, and potatoes. Also recommended for daily consumption are fresh

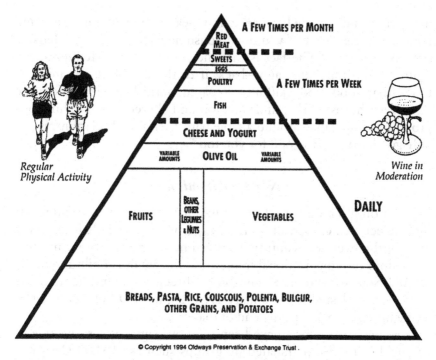

FIGURE 10.1 The Traditional Healthy Mediterranean Diet Pyramid
Courtesy of Oldways Preservation and Trust.

fruits and vegetables, beans and other legumes, and nuts. Olive oil, cheese, and yogurt, in smaller quantities, round out the list of daily diet recommendations. Increasingly more restricted are fish, poultry, eggs, and sweets; at the apex of the pyramid is red meat. Aside from these dietary recommendations, daily exercise and moderate daily consumption of wine *with meals* round out the overall Mediterranean pyramid.

Where does this leave us? In life there are no guarantees; risks and benefits accompany almost all of the decisions we make. Evidence supports claims about the health benefits of wine, providing that it is consumed regularly, at the rate of no more than a glass or two per day, and almost exclusively as an accompaniment to meals. These benefits, however, will accrue primarily to those who consider modifying other elements of diet and lifestyle as well. Longevity can probably be improved with low-fat diets that are high in complex carbohydrates, high in fresh fruits and vegetables, use olive oil rather than butter or margarine, and include little or no red meat; a reduction in stress; and regular exercise (even if you don't consume wine). As Nancy Jenkins (1994:xiii) put it, "The fact is that the people of the Mediterranean figured out a long time ago—back, if truth be told, in the mists of time—that good food, skillfully prepared, garnished with little more than fresh herbs, garlic, and olive oil and shared in something approaching abundance around a table with friends and relations, is not only good tasting; it's good for you too."

Neoprohibitionism

Prohibition has long appealed to a minority of Americans, and it was the subject, remember, of a grand experiment that failed miserably early in the twentieth century. Fostered by religious extremism, then cast within the constitutional framework, the era of Prohibition in the United States (established by the Eighteenth Amendment to the Constitution) lasted from January 17, 1920, until December 5, 1933. Many have yet to accept the end of that era.

In recent years, there have been ominous signs of a rising tide of antialcohol forces, marshaled by groups such as MADD (Mothers Against Drunk Drivers) and SADD (Students Against Drunk Drivers). Aside from drunk driving, other motives behind this new prohibitionist sentiment include alleviating crime and imposing "sin taxes." Furthermore, despite the numerous references to wine in the Bible, many segments of America's religious community have come out strongly against the consumption of alcohol, including wine.

Most of us are against drunk drivers; we are aware of the havoc that they wreak. Most of us are also against drugs and the serious consequences that they have for our society (and keep in mind that they are illegal, even though they are widely available in most parts of the United States).

Wineries and many other representatives of the American alcoholic beverage industry promote the use of designated drivers (someone who is not drinking), responsible drinking (including not driving if you have been drinking), and abstention for those who are inclined toward alcoholism. As noted already, Americans are well aware that alcohol abuse creates serious problems—it has significant economic and social consequences.

Less clear, however, is a way to minimize problems of abuse without taking wine away from all Americans; prohibition will not work.

It nearly destroyed the American wine industry during the 1920s and early 1930s, but alcohol-related problems hardly disappeared. Education about wine consumption would help (and that does not mean throwing it in with cocaine and heroin in drug discussions, as is all too common in school drug programs). A glass of wine with dinner is scarcely comparable to snorting lines of cocaine or mainlining heroin.

Around the Mediterranean, wine is regarded as an integral part of meals, as common on the table as bread and olive oil. On the one hand, even young folks are encouraged to drink a bit of wine (often diluted with water); on the other hand, they are also taught that drunkenness is unacceptable. Americans would do well to think more about what and how we teach children about wine and other alcoholic beverages—perhaps we can find an answer without opting for another era of prohibition.

Changing Demographics and Wine Consumption

The American wine industry is experiencing good times now in the mid-1990s (maybe even a golden age, as some have suggested). However, some in the industry would do well to look at how America is changing and the degree to which demographic changes might affect future wine-consumption trends.

For example, wine in the United States is consumed mainly by middle- and upper-class people of European extraction, and they are going to become an increasingly smaller percentage of the total population. As demographer Carol De Vita (1996:17) recently noted, "The United States has gradually evolved from a largely white, European population to an increasingly diverse society in which people of color and non-European backgrounds represent a growing share of Americans." Nearly one of every four Americans today is either a racial or ethnic minority; one of every eleven Americans today was born in a foreign country (nearly one in every four Californians is now an immigrant). Furthermore, demographers tell us that within another twenty-five years about one-third of all Americans will be minorities, and by the middle of the twenty-first century, Anglos (non-Hispanic whites) will make up only about one-half of the nation's population.

Even within America's minorities, different groups are changing at different rates. In 1995, there were about 70 million minority residents in the country, including 32 million African-Americans and 27 million Latinos. The remaining 11 million were mainly Asians and

Pacific Islanders, and to a smaller extent Native Americans. The Latino and Asian populations, however, have been growing the fastest in recent years because of a combination of high immigration rates and relatively high birth rates (especially for Latinos). Within the next fifteen years, Latinos are expected to outnumber African-Americans. Are these sweeping demographic changes going to affect the American wine industry?

At the same time that notable racial and ethnic changes are occurring, the baby boomers, who began to turn fifty in 1996, are soon going to enter their retirement years. Although they may now be driving the wine market to a considerable degree, that may change with further aging. Behind the baby boomers, the younger so-called Generation X is composed of more minorities and is likely to experience more difficult economic times. Less disposable income, rather than a lack of "taste," may affect how many of these younger people relate to the wine market. As one Generation Xer, Christopher Weir (1996:11), recently put it in an article in *Wine Trader*, "We Xers . . . want a bang for our buck. It means that microbrews are increasingly *earning* our precious dollars with top quality products, that we know a cheap Australian Shiraz blend or Chilean Cabernet Sauvignon is generally of infinitely better character than a comparably priced California wine." Weir went on to suggest that "if the California wine industry continues to alienate the younger generation with its nouveau-riche winespeak, with its reliance on specious studies, and with its dearth of quality in the $4 to $8 range, it can expect a revolution." We might wonder whether the American wine industry is listening.

Competition from Other Beverages

This quotation from Christopher Weir suggests that one of the two changing beverage markets is making claims on the time and money of a growing number of people—the microbrewing industry and its related brewpubs. The other new competition, though certainly less threatening to the wine industry, has arisen through the growth in premium coffee shops, led by Starbucks (of Seattle fame and diffusing rapidly across the nation, even to Japan). According to Kathleen Doheny (1995), the number of "coffee cafés" in the United States is expected to reach 10,000 by 1999.

New Brews

The expansion of microbreweries and brewpubs in recent years has caught almost everyone by surprise, including the major American brewers. The trend began quietly enough, probably with Anchor Steam Beer's revival in the swirling fog of San Francisco. Fritz Maytag purchased Anchor in 1969, then moved it to new headquarters in the former Chase and Sanborn coffee plant on Mariposa Street a decade later. Beginning in the early 1970s, Maytag added some tasty new brews to the original "Steam" beer line, including Anchor Porter, Liberty Ale, Old Foghorn Barley Wine, Anchor Wheat, and an annual Christmas Ale. At about the same time, there were a few other brew kettles heating up, including those of Sierra Nevada in Chico, California, which is now the largest "craft beer" producer in the United States.

By 1985, there were twenty-eight microbreweries in the United States, hardly scratching the surface of America's taste for more flavorful brews than the homogenized products provided by the big brewers. Between 1985 and 1995, however, new brewery doors opened at an accelerating pace; by the end of 1995, there were 717 microbreweries in the nation (275 of which opened in 1995 alone!). Their share of the beer market has risen from a tiny, almost unmeasurable percentage to over 2.0 percent in one decade. New trends in imports have occurred as well, as Canadian microbreweries and Belgian specialty beers have become considerably more popular in the United States.

One of the newest trends among microbreweries is the creation of "fruit brews," made with either fruit extracts or fresh fruit; these have been fashioned after the lambics (fruit ales) of Belgium. Examples include Apple Ale (Niagara Falls Brewing), Georgia Peach (Atlanta's Friends Brewing Company), Samuel Adams Cherry Wheat (Boston Beer Company), Black Dog Honey Raspberry Ale (Spanish Peaks Brewing Company), and a series of flavored ales from Prophet Brewing Company—raspberry, peach, apricot, and passion fruit. Perhaps the most unusual brew so far, however, is Black Chocolate Stout (Brooklyn Brewing Company). Most of these fruity brews are expensive (often $4 or more per 12 oz. bottle), and the degree to which a market for them might develop remains an open question—but they do appeal to Generation Xers, and many are high-quality products.

Another new brew product, the first gender-specific beer, has recently been brought to market by Sydney Goldstein's Golden Pacific Brewery. Known as "Sophie McCall," this brew is distinctive for its extra calcium content (making it the Tums of the brewing world). It is being marketed primarily to middle-aged women.

Although the success of microbreweries would hardly seem threatening in the brewing industry, the big brewers have certainly taken notice. For example, by early 1996 Anheuser-Busch was offering, in addition to its previous lineup, the following new brews: Elk Mountain Amber, Red Wolf, Amber Bock, Zeigin Bock (only in Texas), Faust, Muenchener, Black and Tan, and a seasonal Christmas Brew.

Another measure of interest in these new microbreweries and their quality brews is the appearance of several new publications. Included among them are the following: *American Brewer* (published in Hayward, California), *Brew* (published in Des Moines, Iowa), and *Brewing Techniques* (published in Eugene, Oregon). Other signs of increased interest include the establishment of the Craft Brewing Business Institute at Sonoma State University (Rohnert Park, California); organized festivals and brew tastings, such as the California Festival of Beers (held in Avila Beach, California, where it celebrated its tenth year in 1996 with the help of sixty participating microbreweries); the Micro and Pubbrewers Conference and Trade Show (held in Austin, Texas); and new buying clubs such as Beer Across America (1-800-854-BEER), Hog's Head Beer Cellars (1-800-992-CLUB), and Ale in the Mail (1-800-5-SENDALE).

Premium Coffees

Premium coffee consumption has become a notable trend also, led by Starbucks, which has rapidly expanded from its Seattle origin, first by establishing a mail-order service, then franchising outlets. At the same time, numerous other coffeehouses have appeared as well, offering espresso, cappuccino, latte, and a host of specialty coffees (and often teas as well). Some places include regular poetry readings or jazz sessions; others offer an array of pastries for people on the go. Published in Minneapolis, the *Coffee Journal* provides coffee enthusiasts with up-to-date information about their favorite brews. There is also a new society called the Specialty Coffee Association of America.

Although the availability of fine beers and coffees may in some ways (probably small at this point) compete with wines for consumer dollars, the development of these new markets is probably better read as a sign of broad taste for better-quality products to eat and drink. Unless good wines become prohibitively expensive, wine makers will most likely benefit from customers of the brewpubs and coffeehouses.

Viticultural Problems

As already noted, the 1990s are proving to be a great time for most wineries in the United States, but the wine industry is not without potential problems. Two of the most obvious ones are production problems and pricing (which have been discussed already).

Regional Problems in American Viticulture

Viticultural problems are not the same in each of America's many wine-growing regions. Some producers are probably going to have to depend for a long time on local markets and regional loyalties among consumers. Colorado wines, for example, are not soon going to drive either California or French wines off the shelves of wineshops in Denver, Colorado Springs, or Aspen; they may, however, develop sufficient local interest to allow the state's emerging wine industry to survive. Elsewhere, in Arizona and New Mexico, for example, or in Missouri and Arkansas, the same is probably true. Survival is likely to be tied primarily, if not exclusively, to regional loyalty.

New York and Virginia, by contrast, are producing wines that are becoming important competitors on the national wine scene. Texas seems somewhere in between—Lone Star State pride will undoubtedly play a necessary role in maintaining the industry, but outside investment (from as far away as France) and strong research support from state universities may provide Texas wines with a national niche.

Nonetheless, the western wine regions along the Pacific Rim—California, Oregon, and Washington—will continue to produce the lion's share of American wines in the decades ahead, and there are a few signs of trouble in paradise, especially in the vineyards of northern California. Aside from grape shortages in recent years in the Golden State, which can at least partly be alleviated by better weather and higher yields and by new plantings of popular cultivars (mainly Chardonnay, Merlot, and Cabernet Sauvignon), a grapevine pest and a disease have become significant nuisances.

The Type B phylloxera that has so viciously attacked vines in Napa and Sonoma Counties, especially those planted on AxR#1 rootstock, continues to munch its way from vine to vine. Replanting phylloxera-infested areas has been costly (it can cost up to $25,000 per acre to re-plant a vineyard in the Napa Valley) and time-consuming (because you have to wait for new vines to become productive). A variety of new rootstocks are being used, and rootstock scarcities have some-times slowed replanting efforts.

Nonetheless, in the long run there are some potential benefits from replanting. First, less popular cultivars are being replaced with more popular ones (though popularity can be fleeting, even for grapevines). Second, viticulturists tell us that new trellising and pruning tech-niques are going to bring us not only better vines and more grapes but also higher-quality wines. Even so, many enologists are a bit more wary. In either case, it will be many years before a careful analysis of costs and benefits can be undertaken.

Another problem making an unwanted appearance in parts of both Sonoma County and the Napa Valley is Pierce's disease, which much earlier destroyed the grape industry in Anaheim, California, in the 1880s and in parts of the San Joaquin Valley in the late 1930s. Early signs of the disease include scalding and burning of grapevine leaves in late summer, followed by delayed growth the following spring. Gradually, the plant's root system is destroyed. Sharpshooter leafhop-pers and a few other insects spread the disease. No cure for this bacte-rial disease is known, so attention is focused directly on the carrier, the sharpshooters. Minimizing host environments, such as riparian forests, will help control the numbers of these insects and prevent widespread damage.

Changing Land Uses near Urban Areas

Another problem that viticulturists face is competition from other forms of land use, especially urban uses. The relationship between cities and wine production can be both positive and negative. On the one hand, cities provide ready markets for wine, but on the other hand, as cities grow outward into the surrounding countryside, the demand for land for housing tracts and shopping centers can drive land prices up, forcing agricultural land users to move elsewhere. Cal-ifornia's Santa Clara Valley is perhaps the nation's greatest example of a viticultural area lost to the progress of civilization. Known now as

"Silicon Valley" because of the high concentration of computer and electronics industries in the area, the Santa Clara Valley was once one of the state's premier wine-growing regions, known especially for the quality of its Cabernet Sauvignon. Today, few vines or wineries survive there, and the landscape is almost solidly urbanized; highways, shopping centers, and suburbs stretch out across land that was once covered with grapevines.

In southern California, a similar scene has appeared. During the last three decades, tentacles of the Los Angeles megalopolis have stretched farther eastward, encompassing most of the vineyard land of the Cucamonga Valley, at one time the leading viticultural area in the state. In front of advancing urbanization came both destruction of vineyard productivity from increasing smog and neglect of vines, as farmers awaited lucrative offers from real estate developers.

Anywhere you find vines and urbanization in close proximity, urbanization is apt to be the eventual winner in any struggle over land use, unless there are purposeful attempts (as in the Napa Valley) to protect the land for agricultural purposes.

Imports and Exports

The import and export markets for wine have been changing in recent years; so far, American wineries have been helped more than hurt during the last decade. Between 1985 and 1995, the value of the dollar dropped considerably relative to the currencies of most wine-growing nations (in 1985, for example, $1 was worth almost 9 French francs, but by 1995 it was worth only about 5.5 French francs). Australia has been the only major exception. The lower value of the dollar abroad has the combined effect of making imports to the United States more expensive and exports from the United States less expensive.

Since 1986, the weak dollar has considerably affected the level of imports to the United States from countries such as France and Italy; at the same time, however, imports from Australia have increased because the Australian dollar has remained weak against the American dollar.

As the terms of trade turned against wine imports from most European wine-growing nations, American consumers turned more to American wines. The American wine industry has benefited considerably from what amounts to "protection" from European competition.

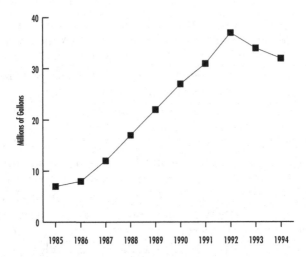

FIGURE 10.2   American Wine Exports, 1985–1994
At least until recently, the American wine industry has benefited from the
weaker dollar.

While holding a comfortable position in the domestic market, American wineries have been able to expand sales abroad, as is apparent in Figure 10.2. Although this has been good for many American wineries, its impact on American wine consumers has been clear—rising prices for American wines, especially for premium varietal wines, and high prices for French and Italian imports as well.

In the long run, the picture can only be described as uncertain—any significant move in the value of the dollar will have an impact, and a sharp upturn in its value, especially against the currencies of France and Italy, would make life difficult for many American wine producers. Right now, however, American wine exports are likely to continue to be strong, though short supplies in recent years and strong domestic demand are going to make it hard for many wineries to supply foreign markets. These combined demand pressures should assure that prices for American wines continue upward, at least for a while. For American wine consumers without large amounts of money to spend, Australian, Chilean, and even South African wines are likely to become more attractive during the remainder of the 1990s.

References

Adams, Leon D. 1986. *The Commonsense Book of Wine*. Rev. and enl. New York: McGraw-Hill Paperbacks.

Adams, Leon D. 1990. *The Wines of America*. 4th ed. New York: McGraw-Hill Book Company.

Amerine, M. A., and Ough, C. S. 1980. *Methods for Analysis of Musts and Wines*. New York: John Wiley and Sons.

Amerine, M. A., Muscatine, Doris, and Thompson, Bob. 1984. *The University of California/Sotheby Book of California Wine*. Berkeley: University of California Press.

Amerine, M. A., and Singleton, V. L. 1977. *Wine: An Introduction*. 2d ed. Berkeley: University of California Press.

Asher, Gerald. 1996. "Up, Up and Away: California's Prestige Sparklers." *Gourmet* 56 (5):54–62.

Augustine, Byron D. 1988. "The Magnificent Mustang." *Mid-South Geographer* 4:34–43.

Automobile Club of Southern California. 1995. *California Winery Tours*. Los Angeles: Automobile Club of Southern California.

Baldy, Marian W. 1993. *The University Wine Course*. San Francisco: Wine Appreciation Guild.

Bank of America. 1973. *California Wine Outlook*. San Francisco: Bank of America NT and SA.

Barsby, Steve. 1995. "The Story Behind the Numbers." *Wines and Vines* 76 (7):16.

Bass, Thomas A. 1986. "The New French Revolution." *Science Digest* 94 (1):61–67, 84–85.

Basu, Janet Else. 1985. "California's Wine Country." *Weatherwise* 34 (2):87–94.

Baxevanis, John J. 1992. *The Wine Regions of America: Geographical Reflections and Appraisals*. Stroudsburg, PA: Vinifera Wine Growers Journal.

Bearden, Bruce E. 1980. "Frost Protection Uses a Variety of Devices." *California Agriculture* 34 (7):38–39.

Beaumont, Stephen. 1995. "Craft Brews Are Here to Stay." *Wine Enthusiast* 8 (12):56–57.

Berger, Dan. 1995. "Chardonnay: Delicacy of Heft." *Los Angeles Times*, October 19.

Berger, Dan. 1996a. "The Scorecard on Reds: U.S. Comes Up Big." *Wine Enthusiast* 9 (1):24.

Berger, Dan. 1996b. "From Soil to Style: Does Terroir Exist?" *Wine Enthusiast* 9 (3):24.

Berger, Dan. 1996c. "Shaking the Vine After the Bug." *Wine Enthusiast* 9 (4):24.

Berger, Dan. 1996d. "Gallo: The New Generation." *Wine Enthusiast* 9 (7):38–41.

Berger, Dan, and Hinkle, Richard Paul. 1991. *Beyond the Grapes: An Inside Look at the Napa Valley*. Wilmington, CA: Atomium Books.

Berkowitz, Natalie. 1996. "Roll Out the Barriques: American Oak Displaces French Oak." *Wine Enthusiast* 9 (1):34–35.

Bernstein, Leonard S. 1982. *The Official Guide to Wine Snobbery*. New York: Quill.

Blue, Anthony Dias. 1980. "Guide to California Cabernets." *Bon Appetit* 25 (9):93–96.

Blue, Anthony Dias. 1981. "California Zinfandel." *Bon Appetit* 26 (2):12, 14.

Blue, Anthony Dias. 1982a. "Sherry." *Bon Appetit* 27 (2):18–19.

Blue, Anthony Dias. 1982b. "Sauvignon Blanc." *Bon Appetit* 27 (7):22–26.

Blue, Anthony Dias. 1983. "Chenin Blanc." *Bon Appetit* 28 (8):18, 100.

Blue, Anthony Dias. 1985. "American Merlot." *Bon Appetit* 30 (10):26, 177.

Blumberg, Robert S., and Hurst, Hannum. 1984. *The Fine Wines of California*. 3d ed. Garden City, NY: Doubleday and Company.

Boulton, Roger B., et al. 1996. *The Principles and Practices of Winemaking*. New York: Chapman and Hall.

Boyd, Gerald D. 1985. "The Wines of Mendocino and Lake Counties." *Wine and Spirits Buying Guide* 4 (3):33–37.

Brenner, Leslie. 1995. *Fear of Wine: An Introductory Guide to the Grape*. New York: Bantam Books.

Broadbent, Michael. 1991. *The Great Vintage Wine Book*. New York: Alfred A. Knopf.

Bullard, Robyn. 1994. "The Power of the Paradox." *Wine Spectator* 18 (20):48.

Cavanaugh, Patrick. 1996. "Pierce's Disease Challenging North Coast." *American Vineyard* 5 (4):4, 26.

Clark, Kenneth. 1969. *Civilisation: A Personal View*. New York: Harper and Row.

Clarke, Oz. 1995. *Wine Atlas: Wines and Wine Regions of the World*. Boston: Little, Brown and Company.

Clay, Grady. 1994. *Real Places: An Unconventional Guide to America's Generic Landscape*. Chicago: University of Chicago Press.

Conzen, Michael P., ed. 1990. *The Making of the American Landscape*. Boston: Unwin Hyman.

Cox, Jeff. 1985. *From Vines to Wines*. Pownal, VT: Storey Communications.

Cox, Jeff. 1995. "Wild Blends." *Wine Enthusiast* 8 (2):28–29.

Crowley, William K. 1984. "United States Viticultural Areas." *Society of Wine Educators Chronicle* (Summer):5–7.

Crowley, William K. 1993. "Changes in the French Winescape." *Geographical Review* 83 (3):252–268.

de Blij, Harm Jan. 1981. *Geography of Viticulture*. Miami, FL: Miami Geographical Society.

de Blij, Harm Jan. 1983. *Wine: A Geographic Appreciation*. Totowa, NJ: Rowman and Allanheld, Publishers.

de Blij, Harm Jan. 1985. "Wine Quality and Climate." *Focus* 35 (2):10–15.

de Blij, Harm Jan. 1986. "Nine Canons of the Geography of Viticulture." *East Lakes Geographer* 21:1–10.

de Blij, Harm Jan. 1988. "The Appeal of Appellations." *Focus* 38 (2):37.

de Blij, Harm Jan. 1991. "America's Zinfandel." *Focus* 1 (1):37.

de Groot, Roy Andries. 1982. *The Wines of California, the Pacific Northwest, and New York*. New York: Summit Books.

De Vita, Carol J. 1996. "The United States at Mid-Decade." *Population Bulletin* 50 (4):1–48.

Diaz, Jo. 1995. "Cat O'Wine Tails." *Wine News* 11 (111):34–35.

Dickenson, J. P., and Salt, J. 1982. "In Vito Veritas: An Introduction to the Geography of Wine." *Progress in Human Geography* 6 (2):159–189.

Dickenson, John. 1990. "Viticultural Geography: An Introduction to the Literature in English." *Journal of Wine Research* 1 (1):5–24.

Doheny, Kathleen. 1995. "Tastes Great, Less Filling?" *Los Angeles Times*, December 19.

Dollar, Tony. 1995. "The Greenhouse Effect." *Wine News* 9 (4):42–43.

Domaine Chandon. 1984. *A User's Guide to Sparkling Wine*. Yountville, CA: Domaine Chandon.

Eliot, T. S. 1963. *Collected Poems: 1909–1962*. New York: Harcourt, Brace and World.

Ensrud, Barbara. 1985. "Santa Barbara Wines." *Cook's Magazine* 6 (1):24.

Ensrud, Barbara. 1988. *American Vineyards*. New York: Stewart, Tabori and Chang.

Ensrud, Barbara. 1995. *Best Wine Buys for $12 and Under: A Guide for the Frugal Connoisseur*. New York: Villard Books.

Farrell, Josh. 1994. "Get Ready for Cal-Ital." *Wine Enthusiast* 7 (3):51–52.

Feiring, Alice. 1995. "Bandol Rouge: A Darker Shade of Rose." *Wine News* 11 (3):28–31.

Findlay, L. M., ed. 1982. *Swinburne: Selected Poems*. Manchester, England: Fyfield Books.

Fleet, Graham H., ed. 1993. *Wine: Microbiology and Biotechnology*. Philadelphia: Harwood Academic Publishers.

Ford, Gene. 1993. *The French Paradox and Drinking for Health*. San Francisco: Wine Appreciation Guild.

Fredericks, Robert. 1969. "Nineteenth-Century Stonework in California's Napa Valley." *California Geographer* 10:39–48.

Galet, Pierre. 1979. *A Practical Ampelography: Grapevine Identification*. Trans. and adapted by Lucie T. Morton. Ithaca, NY, and London: Comstock Publishing Associates.

Gayot, André. 1993. *Guide to the Best Wineries of North America*. Los Angeles: Gault Millau.

Gillette, Paul. 1995. "What's Wrong with This Picture?" *Wine Enthusiast* 8 (9):56–61.

Gregutt, Paul, and Prather, Jeff. 1994. *Northwest Wines: A Pocket Guide to the Wines of Washington, Oregon, and Idaho*. Seattle: Sasquatch Books.

Halliday, James. 1993. *Wine Atlas of California*. New York: Viking.

Hart, John Fraser. 1975. *The Look of the Land*. Englewood Cliffs, NJ: Prentice-Hall.

Hazlitt, W. Carew, ed. 1970. *William Browne: The Whole Works*. New York: Georg Olms Verlag.

Heimoff, Steve. 1995. "Meritage Madness." *Wine Enthusiast* 8 (10):32.

Heimoff, Steve. 1996. "Getting Serious About Sangiovese." *Wine Enthusiast* 9 (3):52–54.

Hinkle, Richard Paul. 1979. *Napa Valley Wine Book*. St. Helena, CA: Vintage Image.

Hinkle, Richard Paul. 1980. *Central Coast Wine Book*. St. Helena, CA: Vintage Image.

Hinkle, Richard Paul. 1986. "Zinfandel: Code Red, but Not Dead." *Wine and Spirits Buying Guide* 5 (2):29–31.

Hiss, Tony. 1990. *The Experience of Place*. New York: Random House.

Holtgrieve, Donald G., and Trevors, James. 1978. *The California Wine Atlas*. Hayward, CA: Ecumene Associates.

Horn, Yvonne Michie. 1995. "Wheeling the Willamette: Touring the Oregon Wine Country." *Wine Enthusiast* 8 (11):38–41.

Hunt, Charles B. 1974. *Natural Regions of the United States and Canada*. San Francisco: W. H. Freeman and Company.

Jackson, John B. 1994. *A Sense of Place, a Sense of Time*. New Haven and London: Yale University Press.

Jackson, Ron S. 1994. *Wine Science: Principles and Applications*. San Diego: Academic Press.

Jenkins, Nancy Harmon. 1994. *The Mediterranean Diet Cookbook: A Delicious Alternative for Lifelong Health*. New York: Bantam Books.

Johnson, Hugh. 1989. *Vintage: The Story of Wine.* New York: Simon and Schuster.

Johnson, Hugh. 1991. *Modern Encyclopedia of Wine.* 3d ed. New York: Simon and Schuster.

Johnson, Hugh. 1994. *The World Atlas of Wine.* 4th ed. New York: Simon and Schuster.

Johnson, Hugh, and Halliday, James. 1992. *The Vintner's Art: How Great Wines Are Made.* New York: Simon and Schuster.

Johnston, Moira. 1979. "Napa, California's Valley of the Vine." *National Geographic* 155 (5):694–717.

Johnston, Moira. 1982. "Beyond Napa." *California* 7 (10):83–87, 136–144.

Kaufman, William I. 1982. *Pocket Encyclopedia of California Wine.* San Francisco: Wine Appreciation Guild.

Kaufman, William I. 1984. *Encyclopedia of American Wine.* Los Angeles: Jeremy P. Tarcher.

Kelley, Denis. 1985. "California's Central Coast." *Wine and Spirits Buying Guide* 6 (6):40–55.

Kliewer, W. Mark. 1980. "Trellising and Spacing Adjust to Modern Needs." *California Agriculture* 34 (7):36–37.

Kramer, Matt. 1995. "Stop Me Before I Judge Again." *Wine Spectator* 20 (8):29.

Kramer, Matt. 1996. "Right Down the Middle." *Wine Spectator* 21 (2):29.

Kunkee, Ralph E., and Cooke, George M. 1980. "100 Years of Wine Microbiology." *California Agriculture* 34 (7):11–13.

Lang, Susan S. 1996. "Will a Drink Improve Your Health?" *Good Housekeeping* 222 (5):72–75.

Laube, James. 1995. *California Wine.* New York: Wine Spectator Press.

Laube, James. 1996a. "Banner Year for Zinfandel." *Wine Spectator* 21 (1):140–149.

Laube, James. 1996b. "Merlot Madness." *Wine Spectator* 21 (7):35–38.

Lee, Wendell C. M. 1994. *U. S. Viticultural Areas.* Rev. 2d ed. San Francisco: Wine Institute.

MacDonald, Kenneth, and Throckmorton, Tom. 1983. *Drink Thy Wine with a Merry Heart.* Ames: Iowa State University Press.

Mansson, Per-Henrik. 1994. "The French Frontier." *Wine Spectator* 18 (20):28–32.

Mariani, John F. 1994. *The Dictionary of American Food and Drink.* New York: Hearst Books.

Massee, William E. 1970. *McCall's Guide to Wines of America.* New York: McCall Publishing.

Matthews, Thomas. 1994. "Wine: Prescription for Good Health." *Wine Spectator* 18 (20):36–44.

Maugh, Thomas H., II. 1996. "Scientists Find Residue of the Ultimate Vintage Wine." *Los Angeles Times,* June 6.

McCarthy, Ed, and Ewing-Mulligan, Mary. 1995. *Wine for Dummies.* Foster City, CA: IDG Books Worldwide.

McConnell, Carol, and McConnell, Malcolm. 1987. *The Mediterranean Diet: Wine, Pasta, Olive Oil, and a Long, Healthy Life.* New York: W. W. Norton and Company.

McNichol, Tom. 1996. "Anchor Aweigh, Full Steam Ahead." *Los Angeles Times Magazine,* March 10.

McPhee, John. 1993. *Assembling California.* New York: Farrar, Straus and Giroux.

Meyer, Justin. 1989. *Plain Talk About Fine Wine.* Santa Barbara, CA: Capra Press.

Miller, Gloria Bley. 1996. *The Gift of Wine: A Straightforward Guide to the Total Wine Experience.* New York: Lyons and Burford.

Mondavi, Robert. Introduction to Justin Meyer, *Plain Talk About Fine Wine.* Santa Barbara, CA: Capra Press, 1989.

Moran, Warren. 1993. "The Wine Appellation as Territory in France and California." *Annals of the Association of American Geographers* 83 (4):694–717.

Morgan, Jeff. 1995. "Touring the Finger Lakes." *Wine Spectator* 20 (9):80–84.

Morgan, Jeff. 1996a. "Boom Time for California Vintners Means Higher Prices and Grape Shortages." *Wine Spectator* 21 (2):10.

Morgan, Jeff. 1996b. "California's 'Other' Coast." *Wine Spectator* 21 (2):48–59.

Moulton, Kirby S. 1980. "Wine—A Multibillion-Dollar Business." *California Agriculture* 34 (7):9–11.

Musto, David F. 1996. "Alcohol in American History." *Scientific American* 274 (4):78–83.

Napa County Department of Agriculture. 1995. *Napa County Agricultural Crop Report: 1994.* Napa, CA: Napa County Department of Agriculture and Weights and Measures.

Newman, James L. 1986. "Vines, Wines, and Regional Identity in the Finger Lakes Region." *Geographical Review* 76 (3):301–316.

Norris, Robert M., and Webb, Robert W. 1990. *Geology of California.* 2d ed. New York: John Wiley and Sons.

Oakeshott, Gordon B. 1971. *California's Changing Landscape: A Guide to the Geology of the State.* New York: McGraw-Hill Book Company.

Olken, Charles, Singer, Earl G., and Roby, Norman S. 1980. *The Connoisseurs' Handbook of California Wines.* New York: Alfred A. Knopf.

Ough, Cornelius S. "Vineyard and Fermentation Practices Affecting Wine." *California Agriculture* 34 (7):17–18.

Page-Roberts, James. 1995. *Wines from a Small Garden*. New York: Abbeville Press.

Parker, Robert M., Jr. 1987. *The Wines of the Rhône Valley and Provence*. New York: Simon and Schuster.

Patton, Dick. 1995. *Life with Wine: A Practical Guide to the Basics*. San Diego: Richard J. Patton Communication.

Perdue, Lewis, Marton, Keith, and Shoemaker, Wells. 1992. *The French Paradox and Beyond: Live Longer with Wine and the Mediterranean Lifestyle.* Sonoma, CA: Renaissance Publishing.

Peters, Gary L. 1984. "Trends in California Viticulture." *Geographical Review* 74 (4):455–467.

Peters, Gary L. 1985. "Viticultural Areas in California." *Long Beach Geographical Notes* 1 (1):4.

Peters, Gary L. 1987. "The Emergence of Regional Cultivar Specialization in California Viticulture." *Professional Geographer* 39:287–297.

Peters, Gary L., and de Blij, Harm J. 1988. "Location, Scale, and Climate: Geographic Dimensions of Viticulture." *Mid-South Geographer* 4:1–16.

Peters, Gary L., and Gossette, Frank. 1990. "Geographic Variations in Cultivar Distributions: Pinot Noir, Barbera and Zinfandel in California." *Journal of Wine Research* 1 (2):121–138.

Peynaud, Emile. 1984. *Knowing and Making Wine*. Trans. from the French by Alan Spencer. New York: John Wiley and Sons.

Pinney, Thomas. 1989. *A History of Wine in America: From the Beginnings to Prohibition*. Berkeley: University of California Press.

Plucknett, Donald L., and Winkelmann, Donald L. 1995. "Technology for Sustainable Agriculture." *Scientific American* 273 (3):182–186.

Robards, Terry. 1980. "Champagne! The Wine of Celebration." *Bon Appetit* 25 (12):173,178, 180.

Robards, Terry. 1996a. "California Cabernets: Can They Age?" *Wine Enthusiast* 9 (5):43–47.

Robards, Terry. 1996b. "Zinfandel: The Mystery Solved." *Wine Enthusiast* 9 (7):44–48.

Robinson, Jancis, ed. 1986. *Vines, Grapes, and Wines: The Wine Drinker's Guide to Grape Varieties*. London: Mitchell Beazley.

Robinson, Jancis. 1994. *The Oxford Companion to Wine*. Oxford: Oxford University Press.

Robinson, Jancis. 1996. *Jancis Robinson's Guide to Wine Grapes*. Oxford: Oxford University Press.

Roby, Norman S., and Olken, Charles E. 1995. *The New Connoisseurs' Handbook of California Wine*. New York: Alfred A. Knopf.

Rollo, Joseph. 1995. "Exports Outpace National Market." *Wines and Vines* 76 (7):18–31.

Rosano, Dick. 1995. "California's Italian Accent." *Wine News* 11 (4):18–23.

Rubin, Hank. 1995. "Champagne: Myths and Facts." *Wine Enthusiast* 8 (11):74–75.

Schaefer, Dennis. 1994. *Vintage Talk: Conversations with California's New Winemakers*. Santa Barbara, CA: Capra Press.

Seldon, Philip. 1996. *The Complete Idiot's Guide to Wine*. New York: Alpha Books.

Sharp, Robert P. 1976. *Field Guide: Southern California*. Rev. ed. Dubuque, IA: Kendall/Hunt Publishing Company.

Sheahan, Randy. 1995. "Bottled Sun: The Wines of Provence." *Quarterly Review of Wines* 17 (4):16–20.

Singleton, Vernon L., et al. 1980. "A Century of Wine and Grape Research." *California Agriculture* 34 (7):4–5.

Smith, Rod. 1996. "Delta Country: The New Force in California Wine." *Wine and Spirits* (June):47–50.

St. John, Bill. 1995. "Where the Bargains Are." *Food and Wine* (September):52–54.

Stanislawski, Dan. 1975. "Dionysus Westward: Early Religion and the Economic Geography of Wine." *Geographical Review* 65 (4):428–444.

Steiman, Harvey. 1995a. "Food for Living." *Wine Spectator* 20 (8):54–59.

Steiman, Harvey. 1995b. "The French Paradox and the Mediterranean Diet." *Wine Spectator* 20 (8):65-67.

Striegler, R. Keith. 1996. "The Changing Varietal Situation in the Southern San Joaquin Valley." *American Vineyard* 5 (4):12–13.

Stuller, Jay, and Martin, Glen. 1994. *Through the Grapevine: The Real Story Behind America's $8 Billion Wine Industry*. New York: HarperCollinsWest.

Sullivan, Charles L. 1982. *Like Modern Edens: Winegrowing in Santa Clara Valley and Santa Cruz Mountains, 1798–1981*. Cupertino, CA: California History Center.

Sullivan, Charles L. 1994. *Napa Wine: A History from Mission Days to Present*. San Francisco: Wine Appreciation Guild.

Sullivan, Valerie. 1996. "New Rootstocks Stop Vineyard Pest for Now." *California Agriculture* 50 (4):7–8.

Sunset Books. 1987. *Wine Country: California*. Menlo Park, CA: Lane Publishing Company.

Teiser, Ruth, and Harroun, Catherine. 1983. *Winemaking in California*. New York: McGraw-Hill Book Company.

Templar, Otis W. 1988. "East Versus West: A Survey of the Texas Wine Growing Industry." *Mid-South Geographer* 4:17–33.

Tish, W. R. 1996. "White Zinfandel Springs Eternal." *Wine Enthusiast* 9 (4):54–55.

Unwin, Tim. 1991. *Wine and the Vine: An Historical Geography of Viticulture and the Wine Trade*. London: Routledge.

Wagner, Philip M. 1973. *A Wine-Grower's Guide*. Rev. ed. New York: Alfred A. Knopf.

Wagner, Philip M. 1974. "Wines, Grape Vines and Climate." *Scientific American* 239 (6):106–115.

Walker, Larry. 1995. "Uncommon Ground." *Wine Enthusiast* 8 (11):52–60.

Weaver, Robert J. 1976. *Grape Growing*. New York: John Wiley and Sons.

Webb, A. Dinsmoor. 1980. "Technology Has Improved Wine Quality." *California Agriculture* 34 (7):6–7.

Weir, Christopher. 1996. "Open Letter to the Wine Industry from a Young Would-Be Curmudgeon." *Wine Trader* Q (1):11.

Whitten, David N. 1995. "The Danish Study: Does Alcohol Matter?" *Wine Spectator* 20 (8):30.

Whitten, David N. 1996a. "Fighting the Puritans." *Wine Spectator* 20 (19):36.

Whitten, David N. 1996b. "The Truth Will Out." *Wine Spectator* 21 (1):33.

Whitten, David N., and Lipp, Martin R. 1994. *To Your Health! Two Physicians Explore the Health Benefits of Wine*. New York: HarperCollinsWest.

Wiegand, Ronn R. 1983. "A Comparison of California and French Sauvignon Blanc." *Wine Country* (July):24–27.

Wine Spectator. 1986. *Wine Maps: The Complete Guide to Wineries, Restaurants, Lodging in California Wine Country*. San Francisco: Wine Spectator.

Wine Spectator. 1996. *Wine Country Guide to California*. New York: M. Shanken Communications.

Winkler, A. J. 1938. "The Effect of Climatic Regions." *Wine Review* 6:14–16.

Winkler, A. J. 1960. "Promising New Areas for Premium Quality Wine Grapes to Replace Acreage Lost to Urbanization." *California Agriculture* 14 (12):2–3.

Winkler, A. J., et al. 1974. *General Viticulture*. Berkeley: University of California Press.

About the Book and Author

Winescapes are unique agricultural landscapes that are shaped by the presence of vineyards, winemaking activities, and the wineries where wines are produced and stored. Where viticulture is successful it transforms the local landscape into a combination of agriculture, industry, and tourism. This book demystifies viticulture in a way that helps the reader understand the environmental and economic conditions necessary in the art and practice of wine making.

Distinctive characteristics of the book include a detailed discussion of more than thirty grape cultivars, an overview of wine regions around the country, and a survey of wine publications and festivals. Peters discusses the major environmental conditions affecting viticulture, especially weather and climate, and outlines the special problems the industry faces from lack of capital, competition, and changing public tastes.

Gary L. Peters, a professional geographer for twenty-five years, is the author of *The Wines and Vines of California*. He lives and teaches in Long Beach, California.

Index

Printed in the United States
by Baker & Taylor Publisher Services